A Modern History of Materials

M. Grant Norton

A Modern History of Materials

From Stability to Sustainability

 Springer

M. Grant Norton
Honors College
Washington State University
Pullman, WA, USA

ISBN 978-3-031-23992-2   ISBN 978-3-031-23990-8   (eBook)
https://doi.org/10.1007/978-3-031-23990-8

This Springer imprint is published by the registered company Springer Nature Switzerland AG
The registered company address is: Gewerbestrasse 11, 6330 Cham, Switzerland

Acknowledgments

I would like to thank the faculty, staff, and students in the Honors College at Washington State University who make each day such a great day to come to work. In particular, a big "thank you" goes to David Shier, Robin Bond, Linda Infranco, and Deb Howe.

Introduction

Since our distant ancestors shaped flint, a polycrystalline sedimentary rock, into rudimentary tools, the majority of human history has been defined by the materials we use—stone, bronze, and iron. But it is over the relatively short period of just 100 years that technological change enabled by innovations in materials science has been particularly rapid—speeding up at unprecedented rates following the patenting of the silicon integrated circuit in 1961. Compare, for instance, the original Motorola StarTAC flip phone from 1996 with a 2022 iPhone 13. They contain many of the same elements in about the same proportions but that is largely where the similarity ends. The transition from the flip phone to the iPhone 13 with all its functionality has happened in a little over two decades. In that time, computers have become exponentially more powerful. In 1996, there were as many as 5 million transistors on an individual integrated circuit. By 2022, that number had exceeded 50 *billion*. Integrated circuits are just one example among numerous others that illustrate the amazing developments in materials, and how we process them, that have happened within the past one hundred years.

Another example of recent rapid advances in materials science is the ubiquitous rechargeable lithium-ion battery, which was commercialized in 1991 enabling an unprecedented, electrified, mobility revolutionizing everything from how we communicate to how we travel. Even as battery performance has increased, with the use of advanced materials, innovations in processing have been able to lower costs to the consumer. Purchasing a lithium-ion battery in 1991 would have cost over $7,500 per kilowatt-hour. Nowadays, a battery with the same power costs less than $180. To put that number into perspective, a Tesla Model S 75D has a 75 kilowatt-hour battery.[1] In 1991, that battery alone would have cost more than half a million dollars!

While completely new materials, like the components of a lithium-ion battery, have been developed over the past century, there are many modern versions of ancient materials, for instance, Corning's Gorilla Glass, which released in 2007 now protects over 8 billion mobile devices. Gorilla Glass shares many of the same chemical constituents—primarily silica—as the Roman glasses produced two thousand years ago. What is distinctly different is the clarity of the glass and the innovative way in which it is processed to overcome the material's inherent low tensile strength.

The 1920s and each of the three decades immediately following were monumental in their far-reaching impact in how we understand, see, and make materials creating a foundation for everything that follows. It is this period, beginning in 1920, that marks the modern age of materials with the awarding of the Nobel Prize in Physics to Swiss metallurgist Charles-Edouard Guillaume. This marked the first time the award had been given to a metallurgist or indeed for a metallurgical discovery.

In ancient times, Guillaume might have been a skilled metalworker smelting, casting, and forging metals to meet a variety of community needs from weapons and tools to domestic items and religious ornaments. Much like his early counterparts working with bronze Guillaume was fascinated by the properties of metals and how those properties could be changed by alloying. The combination that made Guillaume famous was an alloy of iron and nickel named *Invar*, short for *Invar*iable. Invar ushered in an era of what came to be known as "precision metallurgy" where the properties of an alloy could be precisely tailored for a specific application by controlling its composition.[2] With its remarkable stability, Invar would find immediate use in the precise measurement of distance and time. Today, it is manufactured in large-scale quantities for applications that also rely on that stability.

A major development that enabled the rapid technological changes and shift in our relationship to materials over the past century was the creation of quantum mechanics. This new field changed how we *understand* the very structure of materials, in a way that would fundamentally alter how we explain and control their properties. The importance of quantum mechanics is reflected in the number of Nobel Prizes awarded to those scientists most closely associated with its development.

At the end of the twenties, the 1929 Nobel Prize in Physics was awarded to Prince Louis-Victor Pierre Raymond de Broglie for his discovery of "the wave nature of electrons". This was followed closely in the 1930s with Nobel Prizes to Werner Heisenberg for "the creation of quantum mechanics" and to Erwin Schrödinger and Paul Dirac for "new productive forms of atomic theory". Then, in 1945, the Nobel Prize in Physics was awarded to Wolfgang Pauli for his discovery of "a new law of Nature", which determined how the electrons were arranged in an atom.

With the revolutionary quantum mechanical description of the atom, it became possible to explain, for instance, the layered structure adopted by graphite, which acts as the anode electrode of most commercial lithium-ion batteries. Or why one form of carbon—graphite—is a good conductor of electricity while diamond, also comprised of just carbon atoms, is a very poor electrical conductor. Or why silicon solar cells absorb only certain frequencies of the Sun's radiation and how we might go about increasing cell efficiency. Our entire view of the material world since the 1920s is through the lens of quantum mechanics.

Closely following the development of quantum mechanics in the 1920s, which changed how we understand the structure and behavior of materials, was another major innovation in the modern history of materials: the commercialization in the late 1930s and 1940s of the electron microscope. This invention would eventually allow us to clearly see the structure of a material with atomic resolution providing, even in its earliest days, an experimental explanation for many of the properties of materials that had never before been possible.

Since its commercialization, the electron microscope opened an exciting new world beyond what was possible using light microscopy and far beyond what we could see with the naked eye. With its superior resolution, the electron microscope became the essential imaging technique—the workhorse—for observing materials and phenomena at the nanoscale, enabling the eventual development of the field of nanomaterials.

The fourth innovation that defines not only the era of modern materials science but also the modern world itself is our ability to shape materials. Since our ancestors' earliest use of stone tools, we have developed methods to shape materials for our needs. A major breakthrough in our ability to shape materials was the development in the early 1950s of new techniques for growing crystals that would never be found in nature. The fabrication of large, high-purity, defect-free crystals of silicon would become essential for the creation of integrated circuits, the "silicon chips" at the heart of every one of our electronic devices. These are the devices that increasingly impact almost every aspect of our daily lives.

In the 1960s, we saw another innovation in materials processing that would enable a further defining material for our modern world: the glass optical fiber. There are currently over one billion kilometers of optical fiber deployed around the world enabling the transmission of information over long distances, literally at the speed of light.

Each decade during the past 100 years has witnessed not only advances in understanding, synthesizing, and characterizing materials but also the creation of new materials. As an example, the 1930s and 1940s saw the rapid development of a range of materials synthesized from fossil fuel precursors. These were materials with no natural analogs, materials that were purely the result of human ingenuity. The two most widely used—and frequently discarded—of these organic polymers (plastics) are polyethylene, first made in 1933, and polyethylene terephthalate, first made in 1941. While satisfying many of what have become essential applications, polymers along with their chemical precursors are creating an unprecedented array of environmental and human health hazards that were largely unforeseen at the time of their discoveries.

The latter half of the twentieth century and moving into the twenty-first century saw the discovery and development of what have become key innovations in modern materials science: nanomaterials, high-temperature superconductors, lithium-ion batteries, solar cells, and the materials that enable quantum computing. Our relationship with materials impacts almost every aspect of our daily lives, from how we make them, to how we use them, and finally how we dispose of them. That relationship currently looks for the most part like, *take—make—use—dispose*.[3] While this consumerist model may have served well enough for some in the past, it is not sustainable for the future. The visible result of this behavior is obvious with the ever-growing mountains of plastic waste. But the problem extends beyond plastics to include all the metals that we dispose of rather than recycling them at the end of their product's useful life. In the absence of some technological miracle, going forward we will require a more circular economy based on the 4 Rs of sustainability: reduce, retain, recycle, and regenerate.

This modern history explores a wide range of materials and the technologies they enabled. Each of the chapters in this book highlights discoveries and innovations that have occurred within the past century together with some relevant background and context. In each of the chapters, scientists whose important contributions led to the amazing materials described in this book are highlighted. Many of these scientists were awarded Nobel Prizes for their work.

Chapter 1, **A Measure of Stability**, begins this modern history of materials in 1920 with the announcement of the award of the Nobel Prize in Physics to metallurgist Charles-Edouard Guillaume for his discovery of the anomalous properties of the metal alloy *Invar*, a material that defined stability and whose importance has continued to increase over time.

Chapter 2, **A Quantum of Solace**, traces the history of quantum mechanics and the giants of science who gave us this new way of looking at the material world.

In Chapter 3, **Seeing Is Believing**, the focus is on the electron microscope, an instrument critical to our understanding of a whole range of materials. The chapter also provides a very brief history of scanning tunneling microscopy, a technique with the ability to provide atomic resolution images of the surfaces of materials.

Chapter 4, **Made to Measure**, looks at three very different methods of making materials that have become essential to our modern lives—pulling single crystals of silicon, drawing ultra-pure optical fibers, and blowing thin sheets of polyethylene—as well as a fourth method, 3D printing, which offers enormous potential for more sustainable manufacturing processes.

Chapter 5, **There's *Still* Plenty of Room at the Bottom**, follows the history of nanotechnology, from Michael Faraday's experiments at the Royal Society in London on the colors produced by "finely divided" gold particles to Don Eigler at IBM moving a single xenon atom back and forth across a platinum surface. Three nanomaterials—gold nanoparticles, carbon nanotubes, and graphene—are discussed in this chapter, from their discovery to the search for applications.

Chapter 6, **The Future of Mobility**, is about a technology that has been essential in enabling our mobile world: lithium-ion batteries. This chapter describes the history of the lithium-ion battery from the early work in 1913 that demonstrated the potential of a battery containing a lithium electrode up to the awarding of the Nobel Prize in Chemistry to the three scientists who brought the battery into commercial use creating a revolution in portable electronics ushering in a mobile more-electric world.

Chapter 7, **Here Comes the Sun**, discusses solar cells, widely regarded as one of the key technologies necessary for a sustainable, carbon-free, future. Chapter 7 begins by considering the two main approaches used to capture solar energy and convert it into electricity. But the focus of this chapter is photovoltaics. While the first photovoltaic cells were dominated by silicon, recent research has identified other materials that can capture an even greater fraction of the sun's energy for higher efficiency devices.

In Chapter 8, **Certain About Uncertainty**, we look from the materials perspective at the history of the rapidly advancing technology of quantum computing, a direct outcome of the research in the 1920s in quantum mechanics discussed in Chapter 2.

Chapter 9, **Promises Unmet**, discusses several developments in materials science that have produced great excitement in the field but have, as yet, failed to live up to their promise and potential: specifically, carbon nanotubes, introduced in Chapter 5, as well as one of the most exciting discoveries of the twentieth century, high-temperature superconductivity. The chapter concludes with a short discussion of why some of the potential importance of a material or a particular study may be overblown to raise the perceived impact of the work.

Chapter 10, **A Green New Deal**, looks at our history of materials usage and the increasing demand for many critical raw materials. It also addresses some of the challenges in the creation of a circular economy based on the four principles of reduce, retain, recycle, and regenerate. In that context, this chapter looks at some recent studies to upcycle plastics into valuable raw materials that can be used to make new plastics or converted into valuable added chemicals such as waxes and oils. It concludes by examining some of the historically successful approaches to recycling metals and the challenges specifically associated with recycling lithium, the essential element of the lithium-ion battery.

Notes

1. Ritchie H (2021) The price of batteries has declined by 97% in the last three decades. https://ourworldindata.org/battery-price-decline. Accessed 4 Aug 2022.
2. Cahn RW (2001) The Coming of Materials Science. Pergamon, Amsterdam, p 145.
3. Ashby MF (2016) Materials and Sustainable Development. Elsevier, Amsterdam, p 202.

Contents

1

A Measure of Stability

With the Great War in the past, just, for many Americans the 1920s seemed to offer enormous optimism and a great deal to celebrate. It was a time to spend some money, kick-back, and have fun. After a slow start at the beginning of the decade, the stock market was on its way to scaling unprecedented new heights (before, of course, its almighty collapse in October 1929 and the onset of the Great Depression). The economy of the United States was booming with consumers clamoring for the latest technology whether radios to listen to the popular Waldorf-Astoria Orchestra on WJZ Newark or the Eveready Hour, named after the famous battery, or washing machines and refrigerators—the symbols of modern middle-class convenience and comfort. For those who could afford it, there was Henry Ford's Model T, offering a new form of upward mobility for about $300. (Less than $5,000 in today's money.)

The 1920s was a decade of rapid change. Prohibition officially went into effect on January 17, 1920. While reformers rejoiced, organized crime flourished with notorious gangsters such as Al "Scarface" Capone and George "Bugs" Moran capitalizing and profiting from the booming market in illegal alcohol. Women won the right to vote on August 18, 1920, when the 19th Amendment to the United States Constitution was finally ratified. Many politicians including Ohio Republican and 29th President Warren G. Harding pushed for isolationism and there was a major religious revival among conservative Christians. And in a futuristic monoplane named the

M. G. Norton, *A Modern History of Materials*,
https://doi.org/10.1007/978-3-031-23990-8_1

Spirit of St. Louis, Charles Lindberg's solo crossing of the Atlantic Ocean was to usher in the age of modern aviation.

In entertainment and sports, the 1920s were a time of innovation and exceptionalism. Louis Armstrong and the Hot Five were wowing audiences with the new jazz music. Charlie Chaplin was causing confusion and uproar on the silver screen with some of his most famous and popular movies. Ballerina Ana Pavlova was dazzling audiences across the world as *The Dying Swan*. At the 1920 Summer Olympics—the first games since the War—held in the Belgian city of Antwerp, Ethelda Bleibtrey won gold in all three women's swimming contests, on the way breaking five world records. And Babe Ruth was knocking them out of the park at Yankee Stadium and ballparks the length and breadth of the country at the beginning of what would be a fourteen-year record-breaking career with the New York Yankees.

Across the pond, while the immediate post-war years were good for some in Europe there were major challenges the continent had to address with emerging signs of even greater troubles ahead. Britain and France were faced with the crippling debts of war, a deep economic recession, and high unemployment. In Italy, the situation was worse with the economic problems being compounded by the enormous number of casualties the country had suffered coupled with the demoralization brought about by its string of wartime defeats. Collective feelings of disillusionment and despair led the king of Italy to turn to the leader of the Fascist party, Benito Mussolini, to form a government. This decision quickly led to a one-party police state, another sign of trouble ahead.

Germany was on the brink of economic collapse with the requirement to pay reparations to the Allies of 132 billion German marks (roughly $400 billion in today's dollars)—an absolutely staggering amount of money at the time—and the imposed cuts to its industrial and manufacturing base. Russia was engulfed in a vicious civil war that would last until 1922. Whilst Ireland was in the midst of a War of Independence pitting the Irish Republican Army against the British Army and its partners in the Royal Irish Constabulary and the Ulster Special Constabulary, which would lead to a ten-month long civil war.

In Stockholm, the Nobel Foundation was doing what it had done uninterrupted every year since 1901, awarding the Nobel Prize in Physics.

Less well known—at least outside the village of Fleurier—than Louis Armstrong, Charlie Chaplin, Ana Pavlova, Babe Ruth, and possibly Ethelda Bleibtrey, was Swiss scientist Charles-Edouard Guillaume winner of the 1920 Nobel Prize in Physics for, as the Nobel Foundation in Sweden noted, "recognition of the service he has rendered to precision measurements in Physics by

Fig. 1.1 Charles-Edouard Guillaume (ETH-Bibliothek Zürich, Image Archive Portr_08563)

his discovery of anomalies in nickel steel alloys."[1] Awarding the 1920 Nobel Prize in Physics to Charles-Edouard Guillaume, pictured in Fig. 1.1, was the first time the prize had gone to a metallurgist or for a metallurgical discovery. But it certainly would not be the last time a metallurgist, or indeed a materials scientist, or the discovery of a new material would be recognized by the Nobel Foundation.

While many around the world during the 1920s clamored for change, Charles-Edouard Guillaume's career was dedicated to the unchanging—the establishment, reproduction, and delivery of global standards.

Born in Fleurier in 1861 Charles-Edouard Guillaume is certainly the village's most famous son. At the time of his birth, Fleurier—the village of flowers—was best known for watchmaking with almost a quarter of its albeit small population engaged in supplying high quality timepieces to countries as far afield as China, the United States, Egypt, and Turkey. Watchmaking ran in the Guillaume family. Charles' grandfather, Alexander Guillaume, established a successful watchmaking business in London, which he passed down to his three sons. Charles' father relocated the business to Switzerland, settling in the little village of Fleurier in the canton of Neuchâtel.

Guillaume's early education was at the Neuchatel gymnasium or grammar school. Then at seventeen he enrolled in the Zürich Polytechnic, which was later to become the Swiss Federal Institute of Technology in Zürich or ETH Zürich for short. There the young Guillaume studied mathematics. Under the direction of Heinrich Friedrich Weber, Guillaume was awarded his PhD from the University of Zürich in 1883 for a thesis on electrolytic capacitors. (Zürich Polytechnic was not able to award doctorates at that time.)

In 1883 electrolytic capacitors, the topic of Guillaume's dissertation, represented a relatively new technology but one that was important for the emerging electrification of cities around the world and was certainly an area suitable for active research at one of Europe's most prestigious universities. A university that would include not only Guillaume among its famous alumni, but other Nobel Prize winning physicists including Wilhelm Röntgen, the discoverer of X-rays, and Albert Einstein. Einstein, who had been nominated many times, was the favorite to win in 1920, but this again proved not to be his year. Einstein was *awarded* the Nobel Prize in Physics in 1921, the year after Guillaume, for his explanation of the photoelectric effect and *received* his Nobel one year later in 1922. During the selection process in 1921, the Nobel Committee for Physics decided that none of the year's nominations met the criteria as outlined in the will of Alfred Nobel. According to the Foundation's statutes, the Prize can in such cases be reserved until the following year. In Einstein's case it appears that the Nobel Committee did not want to give the award for relativity, which they considered unproven. The compromise was to make the deferred prize for the photoelectric effect, which had been published several years earlier in 1905.

So, what is an electrolytic capacitor, a device that fascinated a young Charles-Edouard Guillaume? Inside an electrolytic capacitor is a metal foil, which acts as the anode or positive electrode. The metal foil is selected for its ready ability to form an oxide layer on its surface, in the same way as rust, iron oxide, forms on a piece of old iron. The earliest electrolytic capacitors used aluminum foil because of the strong bond formed between the oxide layer and the underlying metal.[2] Other suitably oxidizable metals including tantalum and niobium would become increasingly important as the technology developed and more applications were found for these devices.

Tantalum and niobium oxides offered an advantage over aluminum oxide because they are more reliable and can operate stably over a wider range of temperatures. That latter property—stability—was something that would be extremely important to Guillaume later in his professional career.

A feature of all electrolytic capacitors is that because of the presence of the oxide layer formed on the anode, they permit an electric current to flow in one direction only in a process known as rectification. In the late 1880s with the transition from Thomas Edison's low voltage direct current (DC) distribution of electric power to Nikola Tesla's more practical and efficient alternating current (AC), there was a residual need for local low voltage DC electricity. Rectifier function using capacitors could perform the necessary AC to DC conversion improving upon the cumbersome and expensive motor generators that were prevalent at the time.

One example of an important application that required a DC electrical source was to recharge batteries. These were used for, among other things, electric vehicles, which were beginning to appear in the 1800s throughout Europe, in Hungary, the Netherlands, England, France, and in the United States. Although gasoline-fueled internal combustion engines would rapidly come to replace battery power during the nineteenth century, electric vehicles were destined to become an increasingly significant part of the automobile market two centuries later led by companies such as Tesla and the lower profile China-based Kandi. Some recent modeling by the International Monetary Fund predicts that by the early 2040s—midway through the twenty-first century—electric vehicles may represent 90% of all the vehicles on the road.[3] These will all require batteries that can be quickly and conveniently charged and recharged using DC electrical power.

Electrolytic capacitors and their smaller relatives the chip capacitor gained even greater importance with the invention of the transistor, marking the dawn of the electronics age.

The canton of Neuchâtel, which includes the village of Fleurier and the city of Neuchâtel itself has boasted two Nobel Prize winners. In addition to Charles-Edouard Guillaume, who was the first, Daniel Bovet was awarded the 1957 Nobel Prize in Physiology or Medicine for "his discovery relating to synthetic compounds for the blocking of the effects of certain substances occurring in the body, especially in its blood vessels and skeletal muscles."[4] Allergy sufferers are the beneficiaries of Bovet's most well-known discovery, antihistamines. These are used in allergy medications to block the neurotransmitter histamine, which causes the unpleasant runny nose and itchy watery eyes that accompany a mild allergic reaction.

Guillaume's Nobel Prize winning discovery had actually occurred twenty-four years earlier, in 1896, while he was working at the International Bureau of Weights and Measures (*Bureau International des Poids et Mesures*, BIPM) in Sèvres, France, just outside Paris. But the award reflected more than just a single discovery, it recognized the years of important research that Guillaume had performed at the Bureau including establishing and distributing the metric standards.

Located on the left bank of the iconic Seine River, Sèvres already had a strong connection with materials even before the establishment of the BIPM. Since 1756, Sèvres had been well known for the manufacture of elaborately decorated porcelain having established itself along with Meissen in Germany and The Potteries in the United Kingdom as one of the most important sites for the manufacture of European porcelain. Among its notable customers the Sèvres Porcelain Factory welcomed French king Louis XV, who became its principal shareholder and financial backer. Russian empress Catherine II was also a much-valued customer. The czarina's commission was one of the largest and most celebrated ever made by the Sèvres factory. It took multiple iterations before Catherine approved the final design: "After lengthy negotiations about the decorative scheme, the czarina settled on a composition with various Neoclassical elements—stiff bunches of flowers, ample scrolls, and, most important, depictions of cameos, which she collected with enthusiasm."[5]

In 1920, the year Guillaume's Nobel Prize was awarded, Sèvres would again achieve international recognition. This time it was not for its beautiful ceramics or its exacting standards but as the location for the signing of the Treaty of Sèvres between the victorious Allied powers and Turkey, effectively ending the Ottoman Empire. The treaty was signed in the exhibition room at the Sèvres porcelain factory but was never ratified. A treaty described "as brittle as the porcelain that was produced there."[6]

Formation of the International Bureau of Weights and Measures in 1875 in its historic home in the Pavillon de Breteuil coincided with the signing of the Meter Convention (*Convention du Mètre*) or Treaty of the Meter as it is more frequently known in the United States. Seventeen nations, including the United States, were the original signatories. As of January 2023, there are sixty-four Member States and thirty-six Associate States and Economies of the Meter Convention, which forms the basis of all international agreements on units of measurement. The Bureau's task, as home of the International System of Units (SI) and the International Reference Time Scale (UTC), is to ensure worldwide unification of physical measurements.

Guillaume joined the International Bureau of Weights and Measures in 1883 straight after completion of his PhD. His appointment as an assistant researcher came at an important time in the history of the Bureau and for how we measure the world. In 1889, six years after Guillaume had joined, the Bureau embarked on an ambitious program; the approval and worldwide distribution of metric standards. Among his duties, Guillaume was charged with making precise copies of the standard meter, which was kept safely in the Pavillon de Breteuil. The bar was an alloy consisting of 90 parts platinum to 10 parts iridium that had been developed by chemist Henri Sainte-Claire Deville.[7] Both platinum and iridium are highly resistant to corrosion and very stable. The 90/10 alloy had an additional property that was very important. Its dimensions were found to barely change with temperature. If kept outside subjected to the vagaries of the weather the length of the platinum-iridium meter bar would vary by only 0.2 mm between the coldest and warmest average monthly temperatures in Paris.[8] As Guillaume said of the platinum-iridium alloy during his Nobel Lecture on December 11, 1920: "The hardness, permanence, and resistance to chemical agents would be perfect for standards that would have to last for centuries."

Guillaume was correct. Although the original meter bar made of platinum and iridium has lasted well over 100 years it is no longer used to define the metric unit of length. After 70 years, the 1889 platinum-iridium International Meter preserved at the Pavillon de Breteuil lost its position as the primary length standard being replaced by increasingly precise measures to define one meter. Measures that are convenient, can be reproduced, and are unaffected by variables such as temperature. In October 1960 the meter was redefined to an optical standard equivalent to a very precise 1,650,763.73 wavelengths of the orange light, in a vacuum, produced by the element krypton-86 (^{86}Kr).[9] In 1984, the Geneva Conference on Weights and Measures improved upon the definition by stating that a meter is the distance light travels, again in a vacuum, in 1/299,792,458 s with time being measured by a cesium-133 (^{133}Cs) atomic clock.

But back in 1889, Guillaume was faced with the challenge of duplicating the standard meter bar, which because of the amazing stability and the rarity of its constituent elements was incredibly expensive. A single meter cost 7,000 crowns. (At the time a very spacious terraced house for a working London professional, his family, and at least one live-in servant could cost around 400 crowns a year.)[10] It would simply cost too much money to make duplicates for the ever-growing number of member states. If platinum and iridium were too expensive, were there lower cost metals that could be used instead?

Seeking a solution, Guillaume began investigating other alloys that might be used to make duplicates of the standard meter bar.

Nickel satisfied many of the properties necessary to make an unchanging standard. It was "unaffected by the passage of time, rigid and of average expansibility." The challenge turned out to be not in the excellent properties of pure nickel but in finding a factory that could make a bar of the appropriate quality that was "perfectly sound and crack-free."[11] Maybe, instead of pure nickel, an alloy containing nickel might be formulated that had both the desired properties coupled with an ease of manufacture.

Research in iron-nickel alloys was already taking place on both sides of the English Channel (*la Manche*). In Paris, at the request of the Ordnance Technical Department (*Section technique de l'artillerie*) J.R. Benoit was studying the properties of steel alloys containing nickel and chromium with a view to developing a length standard. While in London in the Siemens laboratory at King's College, John Hopkinson was observing some curious magnetic properties of alloys of iron and nickel when they were plunged into solid carbon dioxide (dry ice) at a temperature of $-78°C$.[12]

While interesting, the steel alloys being studied by Benoit and Hopkinson were not suitable for use as standard measures of length. In fact, they were entirely unsuited for this demanding application because they had a high coefficient of thermal expansion (sometimes abbreviated CTE and often represented by the Greek letter α). A high coefficient of thermal expansion means that the dimensions of a material are strongly affected by changes in temperature. By a stroke of good luck, a bar of steel containing 30% nickel—considerably more than was present in the earlier alloys—arrived at the Bureau in Sèvres in 1896. Unexpectedly, Guillaume found that this alloy with its extra amount of nickel had a very low expansivity—one-third lower than that of platinum. Encouraged by this fortuitous result Guillaume continued his studies of iron-nickel (ferronickel) alloys with, as he says, "stubborn obstinacy." He prepared a range of ferronickel alloys with nickel contents from 30% all the way up to 60% then measured how much their dimensions changed with temperature. The composition with the lowest coefficient of thermal expansion contained 64.4% iron and 35.6% nickel. In fact, this alloy exhibited the least amount of thermal expansion of any metal or alloy known at the time. What was particularly surprising to Guillaume about this result was that it did not follow the commonly accepted "rule of mixtures", which would predict that the coefficient of thermal expansion of the alloy should not be less than that of the individual component with the lowest value.

On the suggestion of Marc Thury, a professor at the University of Geneva, Guillaume named this remarkable alloy "Invar", short for *invar*iable. As

Guillaume commented in his 1904 paper published in the journal *Nature*: the succinct name was adopted to avoid saying "steel containing about 36 per cent of nickel, which is characterized by possessing an extremely small coefficient of expansion or by the fact that its specific volume is practically invariable when considered as a function of the temperature."[13] Naming the new alloy Invar was certainly less of a mouthful.

As an invariable measure of length Invar was invaluable. It offered the physical benefits of the platinum-iridium alloy, but at a fraction of the cost. Using Invar, it was now economically feasible for Guillaume to make multiple duplicates of the standard meter bar for distribution throughout the world. Moreover, with its superlative stability it did not take long for Invar to find application in a number of fields in addition to its use as the universal standard of length. One of those was in clockmaking—an application that Guillaume, whose family had a history in the trade, was quick to recognize.[14]

Before the advent of the quartz clock, time was most reliably kept by a pendulum clock measured by the period of swing of a pendulum—a weight (known as a bob) attached to the end of a metal rod. To maintain accurate time keeping it was necessary for the metal rod to always be the same length regardless of temperature (the same requirement for the standard meter bar). If the rod expands the pendulum becomes longer, the clock will lose time. Conversely, a decrease in temperature will cause the clock to gain time. Before the advent of Invar the warming of the steel rods used in pendulum clocks resulted in a loss of ½ second per degree Celsius a day (0.28 s per degree Fahrenheit a day).

To overcome this loss of accuracy, pendulum clocks would employ expansion-compensation mechanisms to reduce the effect of temperature changes. One widely used mechanism was the gridiron pendulum invented by John Harrison in 1721. The gridiron consists of alternating parallel rods of two metals—for instance the two metals might be steel and brass—each having a different coefficient of thermal expansion. Temperature changes are accommodated by the upward expansion in one material (in our example, brass) being compensated with the downward expansion of the other metal (iron). But the complexity of the compensation mechanism was not ideal and a simpler approach was definitely needed. This sounded like a perfect job for Invar, which quickly found its way into pendulum clocks only two years after the discovery of its remarkable properties. The Riefler regulator clock developed in 1898 by Sigmund Riefler was the first clock to use an Invar pendulum giving the timepiece an unprecedented accuracy of 10 ms per day, a "gold standard" in precision. Riefler's new form of pendulum was exhibited at the 1898 World's Fair in Munich. By 1900, Invar pendulum rods were used for

Fig. 1.2 A Riefler clock with its Invar pendulum. This specific clock was purchased in 1904 by the National Bureau of Standards from the firm Clemens Riefler in Germany (Credit NIST)

the highest precision clocks and could be found in nearly every astronomical laboratory. Invar had made the need for complex compensation mechanisms in timepieces redundant. An example of a Riefler clock, purchased by the National Bureau of Standards in 1904, is shown in Fig. 1.2.[15] Riefler clocks served as the national time interval standard until 1929, when it was replaced by the Shortt clock that also used a pendulum and bob made of Invar.

A quick fifteen-minute taxi ride east from Sèvres, home of the International Bureau of Weights and Measures, will take you to Paris's most famous landmark—the Eiffel Tower. Erected in 1889 as a temporary exhibit for the World's Fair, the Eiffel Tower with its 7,300 tons of wrought iron became the tallest structure in the world reaching 1,063 feet at its tip. It would hold that record for over 40 years. Because of the extreme height of the tower the French Service Geographique was concerned about the *sideways* movement of the monument whether caused by wind or by temperature changes, but "for want of an appropriate method no attempt was then made to study the *vertical* movements."[16] By the use of a device made using Invar wire it was possible to measure seasonal changes in the height of the tower. The recording apparatus was quite simple. An Invar wire whose length would imperceptibly change with temperature was connected between the base of the tower and a recording drum located on the second platform, 377 feet from the ground. Over a 24-h period from 8 am on June 8, 1912 to the same time the following day the height of the tower changed by more than 10 mm. The biggest fluctuation was during a downpour that occurred at seven o'clock in the evening causing a sudden drop in temperature with the tower shrinking by a little over 5 mm. During a week-long study at the height of the Parisian summer of 1912 the tower became taller by 20 mm. Changes in the height of the tower tracked very closely with variations in temperature, measured with a thermograph. So, not only is the Eiffel Tower a cultural icon it has also served as a highly sensitive and very large thermometer right in the heart of Paris!

Invar was the first of what became a series of ferronickel alloys with low coefficients of thermal expansion. Although Invar's near zero expansivity was critical for making invariable standards, there were an increasing number of applications emerging for metal alloys that had coefficients of thermal expansion closely *matching* those of various glasses and ceramics. A world changing innovation that brought metals and glass into intimate contact was the incandescent light bulb.

English physicist Joseph Wilson Swan demonstrated a prototype light bulb in 1860 consisting of a carbonized paper filament inside a glass bulb enclosing a vacuum. The filament, which glowed when hot was connected to platinum wires threaded in through the base of the glass envelope that, in turn, made a connection to the holder. Swan's early bulbs produced little light and didn't last long, in part, because of the poor quality of the vacuum that was achievable at the time combined with occasional leaks at the platinum-to-glass joint. Thomas Edison, in his famous patent for the electric lamp granted by the United States Patent Office on January 27, 1880, also described the use of platinum for the lead-in wires forming an air-tight seal with the glass envelope.

Although most ceramics and many glasses expand very little upon heating, even when heated to extremely high temperatures, their coefficients of thermal expansion are rarely zero. In any device, for instance a light bulb that goes through repeated heating and cooling cycles as it is switched on and off, any mismatch in expansivity will create stresses that can lead to vacuum leaks. Eventually the filament will burn out often accompanied by a smoky "pop". While platinum was used as the lead-in wire in early light bulbs because of its closely matched coefficient of thermal expansion with that of the soda-lime-silicate glass used for the bulb and its resistance to oxidation, platinum is far too expensive to be used in a device that was to be manufactured annually in the many hundreds of thousands of units. Guillaume had faced the same challenge—the high cost of platinum—when attempting to duplicate the standard meter bar.

Several manufacturers of incandescent lamps adopted Invar, under the trade name *platinite*, as lead-in wires replacing platinum. In 1904 Guillaume estimated that the use of *platinite* rather than platinum would save several hundred kilograms of the precious and very rare metal. He went on to predict that "if this economy spreads, a ton of platinum may be saved annually for science and those industries in which its use is indispensable."[17] One of the indispensable applications for platinum is in thermocouples for the measurement of temperatures in the range 800–1700°C. Accurate measurement of high temperatures was, and still is, important for scientific studies such as understanding the rate of chemical reactions as well as in controlling the temperatures used in industrial processes such as iron and steel production and glass melting.

In 1911, a new approach was introduced for the lead-in wires for incandescent lamps—the Dumet-wire seal invented by Colin G. Fink a member of the research staff at the General Electric Company Research Laboratories in Schenectady, New York. Dumet wire consists of a ferronickel alloy core with composition 41% nickel and 58% iron with small amounts of manganese and silicon all surrounded by a copper sheath. About a quarter of the total weight of the wire is made up by the copper sheath. The alloy core has a slightly higher nickel content than Guillaume's Invar, 41% compared to 35.6%. Addition of the extra nickel produces an alloy with a coefficient of thermal expansion exactly matched to that of the soda-lime-silicate glass but at a fraction of the cost of platinum. The copper sheath when appropriately oxidized before it is assembled into the light bulb provides a strong and reliable vacuum-tight seal. Copper wire on its own cannot be used for this application because it has a coefficient of thermal expansion much higher than that of the glass bulb. This mismatch would cause tensile forces in the glass during cooling that could break the glass-to-metal seal. Glasses, like most ceramics, are strong when under compression but often very weak in tension.

Although incandescent light bulbs have largely been replaced by compact fluorescent bulbs and more recently LED lighting, Dumet wire is still used in many applications when there is a need to make reliable glass-to-metal seals with soda-lime-silicate glass and lead-containing glasses.

In high intensity discharge lamps, high-powered flashlights, and studio spotlights where the glass bulb is required to withstand intense heat a borosilicate glass is used. Borosilicate glasses have very low coefficients of thermal expansion, about one third that of soda-lime-silicate glasses. Consequently, a new alloy needed to be developed for reliable expansion-matched glass-to-metal seals with borosilicate glasses. That alloy was Kovar—a ferronickel alloy with cobalt.[18] A typical Kovar composition is 54% iron, 29% nickel, and 17% cobalt.

Borosilicate glasses are often preferred over soda-lime-silicate glasses not just for their low coefficient of thermal expansion but because they tend to be stronger, more resistant to chemical attack, and less prone to cracking when rapidly cooled. Pyrex, introduced by Corning in 1915 for laboratory glassware and kitchenware, is a familiar everyday example of a borosilicate glass. Many of us may be well acquainted with the excellent shock resistance of Pyrex!

Borosilicate glass was also the preferred choice to provide hermetic seals for early transistors. A reliability problem with these newly invented devices was water vapor in the atmosphere penetrating to the transistor junction. Vacuum-tight encapsulation of the transistor was essential to attain high reliability. The 2N559 shown in Fig. 1.3 is a germanium switching transistor made in the late 1950s by Western Electric in Laureldale, Pennsylvania for use by the military in its Nike Zeus anti-ballistic missile system. Three Kovar-to-glass seals were used to bring the three external wires into the top hat shaped package containing the transistor. With its stable seals the 2N559 was one of the most reliable transistors of the time. It had an impressively low failure rate of about 0.001% per 1,000 h or only 10 failures in 1 billion hours.[19] By mid-1958, Western Electric's germanium transistors of the 2N559 type were circling the Earth in *Explorer* and *Vanguard* satellites as the United States and Russia competed in the Space Race. By the end of the 1950s, germanium transistors were making their way into high-quality, inexpensive transistorized products such as miniaturized hearing aids and solid-state radios.

Guillaume's discovery of the anomalous properties of Invar, the invariable alloy whose dimensions would barely change when the metal was heated or cooled, had almost immediate impact on defining distance and accurately measuring time. Invar formed the basis for a number of ferronickel alloys whose expansivity could be matched to that of glass allowing bright long-lasting incandescent lights and reliable transistors. One hundred years after Guillaume was recognized for his discovery, Invar, has continued to grow in importance. Its applications exploit the unusual property that Guillaume first observed in 1896—its stability.

Fig. 1.3 Invar glass-to-metal seals were important in the early electronics industry. These are two examples of a 2N559 germanium switching transistor originally released in 1957 by Western Electric. Early devices were packaged in a gold-plated metal can. Later 2N559s were packaged in black painted metal cans (Reprinted with permission Industrial Alchemy)

While a considerable amount of research effort nowadays is in discovering materials that have large measurable responses to a stimulus, whether that be temperature, pressure, an applied force, the presence of a specific chemical or biomarker, there remain many applications where stability is critical. Invar's legendary stability has made it an essential material for delivering energy whether that is in the form of Invar-lined shipping containers that are used to transport liquid natural gas or as the core of aluminum-clad overhead power cables that carry electricity from power stations to our homes and businesses.

An application Guillaume could not have foreseen was the use of Invar for the enormous metal tanks used to ship liquified natural gas (LNG) around the world.

The first tanker shipment of liquified natural gas took place from Lake Charles, Louisiana bound for Canvey Island in the Thames Estuary in England in 1958 aboard the experimental vessel the *Methane Pioneer*.[20] Natural gas is a vital energy source for countries around the world including the United States where it is the primary heating fuel for almost half of all American households. Most natural gas is delivered by pipelines, but the growth in the global market has required it to be transported to places pipelines either don't exist or can't reach. When cooled to around $-160°C$ the gas liquifies, which decreases its volume allowing it to be transported via ocean going tanker ships. In 2020, the United States exported 2,390 billion cubic feet of liquified natural gas to forty countries on five continents.[21] When the liquified natural gas reaches its destination it is warmed back to its gaseous state for distribution to consumers and power plants.

Invar sheet about 0.5 mm thick is one of the materials used in the construction of the tanks—especially a type of tank called a membrane tank where the ship's hull is the tank's outer wall. The Invar forms the inner insulating barriers containing the cold liquid. Because of its very low coefficient of thermal expansion, Invar has incredible stability over a wide range of temperatures even down to the extremely frigid temperatures of liquified natural gas. Another advantage of Invar for this application is that because the dimensions of the alloy don't change significantly when heated or cooled it is not necessary to make corrugations in the tank structure to accommodate dimensional changes. As a result of using Invar less material is required in the manufacture of the tanks.

Invar is not only important for moving liquified natural gas around the world it also assists in efficiently moving electrons over long distances. The overhead transmission lines that we see hanging from steel towers transmit high-voltage electricity from hydroelectric, thermal, or nuclear plants to local substations, which then distribute the power to our homes and businesses. Invar wire clad with aluminum enables an increased transmission capacity with lower losses when compared with conventional cables. Consequently, lowering the large carbon footprint associated with electricity transmission. Invar not only provides stability during changes in the temperature of the cables because of its very low coefficient of thermal expansion it also prevents sagging of the cable when it gets hot.

While we know experimentally the unusual properties of Invar, the most important being its very low expansivity when heated, a detailed mechanism of what has been termed the "Invar effect" or sometimes the "Invar problem" that explains the behavior of Invar has eluded scientists for over 100 years.[22,23] A discovery in 2001 made by a collaboration involving Uppsala University and the Royal Institute of Technology in Sweden, the European Synchrotron Radiation Facility in Grenoble, France, and Sandia National Laboratories in the United States provided some new insights into the "Invar effect" when it was shown for the first time that pressure could transform a "normal" metal alloy, one that expanded normally at atmospheric pressure, into an "Invar" material.[24] This observation required the use of high-pressure diamond anvil cells to compress the metal and powerful beams of X-ray to probe infinitesimal changes in the material's structure. A further observation that has added complexity to the more than century-old struggle to comprehend the microscopic origin of Invar behavior is that it has been found in alloys that don't even contain iron or nickel. For instance, researchers at Los Alamos National Laboratory and Argonne National Laboratory have observed the Invar effect in plutonium–gallium alloys.[25] This is a very different combination of elements than found in Invar.

Foreshadowing some of the challenges we are acutely aware of today, Charles-Edouard Guillaume was concerned about the sustainability of our supply of rare metals. He was, for instance, keen to see Invar replacing platinum as the lead-in wires in the rapidly growing incandescent lamp industry. The raw materials used to make Invar are considerably more abundant than platinum, which occurs in the Earth's crust at just one-millionth of one percent, making it one of the rarest of all elements. In Guillaume's view the limited supply of platinum should be saved for those uses where it was irreplaceable.

While Charles-Edouard Guillaume's work was concerned with stability, precision, and certainty the next chapter describes the development of a theory based on uncertainty that would revolutionize our understanding of materials and the atoms from which they are made.

Notes

1. Charles Edouard Guillaume—Facts. NobelPrize.org. Nobel Media AB 2020. Sun. 25 Oct 2020. https://www.nobelprize.org/prizes/physics/1920/guillaume/facts/.
2. Pollak C (1896) Elektrischer flüssigkeitskondensator mit aluminiumelektroden. German Patent Nr. 92564. First patent for the electrolytic capacitor.
3. Cherif R, Hasanov F, Pande A (2017) Riding the energy transition: oil beyond 2040. International Monetary Fund, Working Paper No. 17/120. Available at https://www.imf.org/en/Publications/WP/Issues/2017/05/22/Riding-the-Energy-Transition-Oil-Beyond-2040-44932. Accessed 5 Aug 2022.
4. Daniel Bovet—Biographical. NobelPrize.org. Nobel Prize Outreach AB 2021. Sun. 22 Aug 2021. https://www.nobelprize.org/prizes/medicine/1957/bovet/biographical/.
5. Préaud T (1995) The Sèvres porcelain service of Catherine II of Russia: the truth concerning payment. Studies in the Decorative Arts 2:48–54.
6. Karcic H (2020) Sèvres at 100: The peace treaty that partitioned the Ottoman Empire. Journal of Muslim Minority Affairs 40:470–479.
7. Among his many accomplishments in the field of materials, Henri Sainte-Claire Deville developed a process to obtain metallic aluminum and a method for manufacturing sodium. Together with Friedrich Wöhler, Deville discovered the ceramic material silicon nitride.

8. The coefficient of linear thermal expansion (CTE) for the Pt90-10Ir alloy is 8.7×10^{-6} K^{-1} between 20 and 100 °C. Average monthly temperatures in Paris vary from 2°C to 25°C.

9. Page CH, Vigoureux P (1975) The international bureau of weights and measures 1875–1975. National Bureau of Standards Special Publication 420:1–256.

10. Emsley C, Hitchcock T, Shoemaker R London history—currency, coinage and the cost of living. Old Bailey Proceedings Online. www.oldbaileyonline.org, version 7.0. Accessed 24 Aug 2021.

11. Guillaume CE (1920) Invar and Elinvar. Nobel Lecture, December 11. The text of Guillaume's lecture is available at https://www.nobelprize.org/uploads/2018/06/guillaume-lecture.pdf. Accessed 5 Aug 2022.

12. Hopkinson J (1890) Magnetic properties of alloys of nickel and iron. Proceedings of the royal society. This is the first of a series of publications on this alloy type. Available at https://royalsocietypublishing.org/doi/pdf/10.1098/rspl.1890.0001. Accessed 5 Aug 2022.

13. Guillaume CE (1904) Invar and its applications. Nature 71:134.

14. Guillaume CE (1904) Invar and its applications. Nature 71:137.

15. In 1988 the National Bureau of Standards was renamed the National Institute of Standards and Technology (NIST).

16. Guillaume CE (1919) The anomaly of the nickel-steels. Proceedings of the Physical Society of London 32:375.

17. Guillaume CE (1904) Invar and its applications. Nature 71:139.

18. Kover is a trademark of Carpenter Technology.

19. Smits FM (ed) (1985) A history of engineering and science in the bell system. AT&T Bell Laboratories: Indianapolis, p 54.

20. Jensen JT (2004) The development of a global LNG market: is it likely? If so, when? NG 5 Oxford Institute for Energy Studies, p 8. Available at https://www.oxfordenergy.org/wpcms/wp-content/uploads/2010/11/NG5-TheDevelopmentofAGlobalLNGMarketIsItLikelyIfSoWhen-JamesJensen-2004.pdf. Accessed 5 Aug 2022.

21. United States Department of Energy, Liquified Natural Gas (LNG). Available at https://www.energy.gov/fe/science-innovation/oil-gas/liquefied-natural-gas. Accessed 5 Aug 2022.

22. Lagarec K, Rancourt DG, Bose SK, Sanval B, Dunlap, RA (2001). Observation of a composition-controlled high-moment/low-moment transition in the face centered cubic Fe–Ni system: Invar effect is an expansion, not a contraction. Journal of Magnetism and Magnetic Materials 236:107–130.

23. van Schilfgaarde M, Abriksov IA, Johansson B (1999) Origin of the Invar effect in iron-nickel alloys. Nature 400:46.

24. Dubrovinsky L, Dubrovinskaia N, Abrikosov IA, Vennström M, Westman F, Carlson S, van Schilfgaarde M, Johansson B (2001) Pressure-induced Invar effect in Fe–Ni alloys. Physical Review Letters 86:4851.

2

A Quantum of Solace

The first Nobel Prize in Physics awarded at the beginning of the Roaring Twenties was in recognition of stability, of constancy, of certainty. The last Prize of the decade was for one in a rapid series of groundbreaking ideas that reflected uncertainty and, to many, an uncomfortable lack of precision. In the ten years following the award of the 1920 Nobel Prize in Physics to Charles-Edouard Guillaume, a metallurgist from the small Swiss watchmaking town of Fleurier, a scientific revolution would take place transforming our understanding of the structure of matter. We would now be able to explain, for instance, why the arrangement of the elements in the Periodic Table made sense, why atoms join together in specific ways, and how those bonds between atoms determine many of the properties of materials.

Closing out the decade, the 1929 Nobel Prize in Physics was awarded to Prince Louis-Victor Pierre Raymond de Broglie "for his discovery of the wave nature of electrons."[1] Five years earlier de Broglie, while a member of the Faculty of Sciences at Paris University, introduced the idea that all matter, most importantly electrons, could be described not only as particles but also as waves. de Broglie's graduate thesis titled *Recherches sur la théorie des Quanta* (Researches on the quantum theory) gained him a doctorate while his hypothesis laid the foundation for an entirely new theory called quantum mechanics.

While it was widely accepted by the beginning of the twentieth century that light could be regarded as having both particle and wavelike properties, that extension had not been made to electrons until de Broglie's hypothesis.

© The Author(s), under exclusive license to Springer Nature
Switzerland AG 2023
M. G. Norton, *A Modern History of Materials*,
https://doi.org/10.1007/978-3-031-23990-8_2

Since the announcement of the discovery of the electron by Joseph John "J.J." Thomson in 1897, during a Friday evening lecture to the Royal Institution in London, its nature had always been regarded as being that of a particle: a negatively charged particle. As Thomson himself described what were known in the field at the time as cathode rays: "... another view of these rays is that, so far from being wholly aethereal, they are in fact wholly material, and that they mark the paths of particles of matter charged with negative electricity."[2] Thomson was awarded the 1906 Nobel Prize in Physics "in recognition of the great merits of his theoretical and experimental investigations on the conduction of electricity by gases."[3] He not only showed that cathode rays consist of particles—electrons—that conduct electricity, he also concluded that electrons are part of atoms.

The structure of the atom, and specifically the arrangement of electrons in an atom, was a question that was taxing some of the greatest scientific minds in the early twentieth century. While a young Louis de Broglie was completing a science degree in 1913 before being conscripted for military service, Danish physicist Niels Bohr proposed his "planetary" model of the atom. This model was the first that satisfactorily described some of the experimental observations being made at the time. Bohr's Nobel Prize in Physics was awarded in 1922 for "services in the investigation of the structure of atoms and of the radiation emanating from them."[4]

The planetary model was a radical deviation from the popularly nicknamed "plum pudding" model that was proposed by Cambridge's J.J Thomson.[5] Thomson imagined the negatively charged electrons that he had recently discovered dotted throughout a positively charged ball. The number of positive charges was equally balanced by the number of electrons. In, perhaps, a typically British P.G. Wodehouse style description Thomson's model of the atom was thought to resemble a Christmas pudding, a sticky ball of sugar and flour randomly interspersed with currants.

Bohr's model separated out the nuclear pudding from the much lighter electrons. These were then placed in defined orbits around the nucleus, which is itself comprised of a combination of protons and neutrons. The nucleus is tightly bound and heavy, much like many plum puddings! And the electron orbits are far removed—in atomic terms—from the nucleus. This aspect of Bohr's model fit nicely with the results of experiments being done by New Zealander Ernest Rutherford, an experimental physicist working at the University of Manchester in the north of England. Under Rutherford's direction his senior laboratory assistant Hans Geiger together with undergraduate student Ernest Marsden fired alpha particles (the nuclei of helium atoms) at thin gold foils to observe what happened to their paths. A radioactive radium source held in a lead block was the source of the alpha particles. What Geiger and Marsden found was that most of the alpha particles went straight through the foil as though it wasn't even there. After all, it was thinner than the width of a human hair. A smaller number of the alpha particles were scattered in random directions. Some were even reflected back towards the source itself. This result was entirely unexpected. Rutherford described it as "quite the most incredible event that has ever happened to me in my life. It was almost as incredible as if you fired a 15-inch shell at a piece of tissue paper and it came back and hit you."[6]

The only reasonable interpretation of these scattering results was that the nucleus must be very small with most of the atomic volume being taken up by the orbiting electrons. A typical nucleus is only 10^{-15} m wide, whereas the diameter of an atom accounting for the surrounding electrons is 100,000 times larger at about 10^{-10} m. (A length of 10^{-10} m is equivalent to 1 Ångstrom; an old unit but often used to express very small distances.)

Ernest Rutherford's Nobel Prize, which was in Chemistry rather than Physics, was awarded in 1908 "for investigations into the disintegration of the elements, and the chemistry of radioactive substances."[7] What Rutherford had discovered were two types of radiation. He named them "alpha rays" and "beta rays". The alpha rays were soon shown to be the nuclei of helium atoms consisting of two protons and two neutrons. Beta rays are electrons. Rutherford's experiments with the very able Geiger and Marsden on the structure of the atom were probably more significant and arguably had a greater impact within the scientific community than the work that had won him the Nobel Prize in Chemistry, but he did not get a second in Physics. So far only Marie Curie has received that honor. Radium and polonium, the two elements discovered by Curie are both alpha particle emitters.

An essential feature of Bohr's atomic model was its requirement that each electron circled the nucleus at a fixed distance, in the same way that the planets in our Solar System orbit the Sun. We know exactly where each planet is at any moment in time. And the closer ones we can often pick out predictably on a clear evening. In Bohr's model, an electron's position and energy are both known with certainty—just like that of an orbiting planet. If electrons went from one orbital to another (either one closer to or further away from the nucleus) the atom would either radiate or absorb energy. Absorption of a specific and measurable amount of energy would move an electron into a more distant orbit. On the other hand, energy is radiated by the atom when that electron returns to its original "ground" state. Because the electrons can only have specific energies the orbits are said to be *quantized*.

The energy difference between quantized orbits is unique to each atom. For instance, in our description of the meter standard in Chapter 1, one definition was that it is 1,650,763.73 vacuum wavelengths of the orange light produced by the element krypton-86. That wavelength corresponds to a precise 605.7802106 nm and is emitted when an electron in an atom of krypton moves from an outer orbital of higher energy to an inner orbital of much lower energy.[8] Australian researchers C.F. Bruce and R.M. Hill, both at the Division of Metrology at CSIRO in New South Wales, measured the wavelength of the orange light from krypton-86 (^{86}Kr). They also measured the vacuum wavelengths of the two isotopes mercury-198 (^{198}Hg) and cadmium-114 (^{114}Cd). The value they obtained for the mercury isotope was 546.2 nm, corresponding to emission of green light. Cadmium radiated in the red with a longer wavelength of 644.0 nm. These emissions correspond to unique electron transitions within the atoms of mercury and cadmium, which is consistent with Bohr's atomic model.

So, while gaining broad acceptance, Bohr's model was not complete because while it was consistent with Rutherford's observations it was unable to successfully explain other experimental results. For instance, the model could not explain why the alkali metals sodium and potassium are so alike in many of their physical and chemical properties. They are both soft silvery-white metals that react violently with water. Or why fluorine and chlorine are so similar in their ability to react with the alkali metals to form salts that readily dissolve in water. It also, even with some subsequent improvements by Bohr, could not produce the correct arrangement of electrons in the noble gas atoms, such as neon, argon, and krypton.[9] These are important expectations for a realistic atomic model.

Considering the electron as a tiny planet following a well-defined path around its nucleus-centered solar system was abandoned for a quantum mechanical approach, where electrons while having the properties of a particle also behaved like waves. This was the hypothesis that came from de Broglie's doctoral research in Paris: electrons, like light, could behave as both a particle and as a wave.

Albert Einstein had proposed that light could be considered to be both a particle and a wave in order to explain two experimental observations—the photoelectric effect that was first demonstrated by Philipp von Lenard in 1902 and the results of Thomas Young's double slit experiment, which were reported to the Royal Society in 1803, almost one hundred years before von Lenard's research.

The photoelectric effect is the release of electrons from a material when it is exposed to light of a certain frequency. In his experiments, von Lenard shone ultraviolet light onto a clean metal plate inside a vacuum tube, which caused it to generate electrons. These he could manipulate using applied magnetic and electric fields. What was unexpected about von Lenard's study was that when the speed of the ejected electrons was measured it was found to be "independent of the ultraviolet light intensity."[10] It would not be unreasonable to expect that a brighter light might make the electrons move faster. They might have a greater energy. But increasing the intensity of the incident light only produced more electrons, each with the same energy. Further experiments showed that using light with a higher frequency (a shorter wavelength) produced more energetic electrons. If the light had a sufficiently low frequency, then no electrons at all might be ejected. Philipp von Lenard was awarded the 1905 Nobel Prize in Physics "for his work on cathode rays (electrons)", which included his studies of the photoelectric effect.[11]

In 1905 Einstein explained the photoelectric effect, which had been observed by von Lenard by thinking of light in the form of packets of energy, or quanta. American physical chemist at the University of California Berkeley Gilbert Newton Lewis named these units of light "photons" in a 1926 letter to the journal *Nature*.[12] When a photon struck an atom in von Lenard's metal plate there was a transfer of energy causing an electron—a photoelectron—to be ejected. Higher frequency radiation, for instance ultraviolet light, contains higher energy photons than lower frequency visible light or infra-red radiation allowing ejection of higher energy electrons. A brighter light simply contains more photons, but they all have the same energy. The converse is true for fainter light.

While the photoelectric effect could be satisfactorily explained by considering the particle nature of light, Thomas Young's double-slit experiment had already proved beyond any reasonable doubt the wavelike behavior of light.

Thomas Young was an English polymath who made significant contributions to many fields from Egyptology to medicine and from music to solid mechanics. His name is permanently associated with the elastic behavior of materials, which is described by a parameter known as Young's modulus, E: the simple ratio of applied stress over strain. Materials with a large Young's modulus resist being elastically deformed. Most ceramics, for instance porcelain, have a high modulus indicating that they deform very little before breaking. Rubber on the other hand has a very low Young's modulus, which reflects the ease with which we can, for example, stretch a rubber band a very long way and still have it return to its original shape when we let go. Young's equation, later named the Young-Dupré equation, defines the conditions for when a liquid will wet and spread on a solid surface. And the Young–Laplace equation relates to capillary action—how far liquid rises within a narrow tube.

Thomas Young also contributed to the deciphering of Egyptian hieroglyphs including the Rosetta stone, "Young's Rule" in medicine was a means for converting adult doses of a medication into those suitable for children. Young's temperament, which he described in 1800, is used in tuning musical instruments. But in Thomas Young's own judgement among all his many achievements the most important was to establish the wave theory of light.

In the 1790s, while at Cambridge University in England, Thomas Young experimented with sound waves investigating how they propagated through air. He was particularly interested in the phenomenon of interference where individual waves could sum to make a louder sound or cancel each other out, the principle behind today's noise-cancelling headphones. Young also studied how waves in water would interfere when they passed through holes made in a barrier. When he used two holes the incident waves would pass through each hole spreading out in a semicircular pattern. Eventually, the waves would intersect causing some to become larger where the peaks of one ripple combined with the peak of another. In regions where a peak overlapped with a trough in another ripple the resulting wave would be smaller. Young observed "a magnificent example of interference of two immense waves with each other; the spring tide being the joint result of their combination [of solar and lunar tides] when they combine in time and place, and the neap tide when they succeed each other at the distance of half an interval, and so as to leave the effect of their difference only sensible."[13]

Experimenting with sound and water was a precursor to the famous "Young's double-slit experiment" with light.

The experiment was beautifully simple, yet the implications profound. As Young put it when presenting his results to the Royal Society in London in 1803: "The experiments I am about to relate … may be repeated with great ease, whenever the sun shines, and without any other apparatus than is at hand to every one."[14] A light source in a darkened room was shone on a series of cards. The first card had a single aperture in the form of either a tiny pinhole or a slit. The second of these cards had a pair of holes or slits. Finally, there was a blank screen where the light that had passed through both preceding cards arrived. Young observed that the light arriving at this final screen produced a pattern of alternating light-and-dark stripes, an interference pattern exactly equivalent to the interference pattern of water waves observed in the earlier ripple-tank experiment.

Together, the double-slit experiment and Einstein's explanation of the photoelectric effect that light consists of quanta—photons—that must have a certain minimum energy before they can liberate an electron showed that the concept of wave-particle duality applied to light. If duality was equally to apply to electrons, as hypothesized by de Broglie in 1924, then they too would have to show wavelike properties.

Experimental proof that electrons could behave as waves came in 1927 with the publication of the research results of Americans Clinton Davisson and Lester Germer, scientists at Bell Telephone Laboratories (Bell Labs) in New York.[15] Conceptually the experiment, which was set up towards the end of 1926 was, like Young's, very simple. A source of electrons coming from a heated tungsten ribbon placed within a vacuum were accelerated using electric fields towards a target comprising a single crystal of nickel. The target could be rotated to different angles and the electron detector, which was used to monitor the scattered electrons, could also be moved in order to observe the directions that the electrons were coming back from the metal surface. What Davisson and Germer found, to their great surprise, were distinct peaks in the intensity of the scattered electron beam that occurred only at certain specific angles. The position of these peaks was found to be consistent with Bragg's Law, which had been derived in 1912 to describe the diffraction of X-rays, high energy electromagnetic radiation, by a crystal. William Henry Bragg and his son William Lawrence Bragg had jointly developed the eponymous law and shared the 1915 Nobel Prize in Physics "for their services in the analysis of crystal structure by means of X-rays."[16]

The electron scattering results obtained by Davisson and Germer not only were consistent with Bragg's Law they were experimental support for the prediction of Louis de Broglie that electrons behaved like waves.

At the same time that Davisson and Germer were working on demonstrating the wavelike behavior of electrons in the United States, British physicist George Paget Thomson (the son of J.J. Thomson) together with his student Alexander Reid were conducting similar experiments at the University of Aberdeen in Scotland. Rather than using a relatively thick crystal of nickel as the target, as Davisson and Germer did, Thomson and Reid's initial experiments used a thin film of celluloid—a plastic. Estimated by Thomson to be about 50 nm thick the film was thin enough that it was transparent to the electron beam. Any electrons that passed through the celluloid film struck a photographic plate producing a pattern consisting of a bright central spot together with a single clearly visible concentric ring—a diffraction pattern that would only be possible if the electrons were behaving as waves, as would be consistent with de Broglie's hypothesis. Paget Thomson working alone confirmed and extended his experimentation to films of gold, aluminum, (the metals were less than 100 nm thick, still just about see-through to electrons)

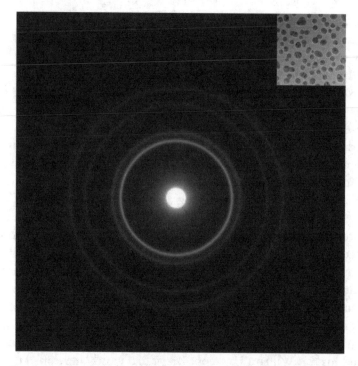

Fig. 2.1 Proof that electrons are waves. An electron diffraction pattern, obtained with a modern electron microscope, from gold nanoparticles on a carbon film. The inset shows the gold nanoparticles, which are in darker contrast than the carbon film (Reprinted with permission Gatan)

and "of an unknown (at first thought to be platinum, but probably "some greasy" organic) substance."[17]

Figure 2.1 shows an example of an electron diffraction pattern from a sample consisting of gold nanoparticles on a carbon film, which is very similar to the one obtained by Paget Thomson. In addition to the bright central spot there are a series of continuous concentric rings where the diffracted electrons are in phase. These rings become progressively fainter the further away they are from the central spot. Destructive interference is responsible for the dark regions between the rings. By measuring the diameter of the rings and using the equation developed by the Braggs, Thomson was able to determine that the size of the unit cell of gold was 0.380 nm on a side.[18] The value obtained from X-ray diffraction measurements was 0.4065 nm—a discrepancy of only 6½ percent.

Two subtly different experiments had almost simultaneously proved that de Broglie was right. Or, as Clinton Davisson put it in his Nobel Lecture: "That streams of electrons possess the properties of beams of waves was discovered early in 1927 in a large industrial laboratory in the midst of a great city, and in a small university laboratory overlooking a cold and desolate sea … Discoveries in physics are made when the time for making them is ripe, and not before."

Almost a decade after de Broglie had received the Noble Prize in Physics, Davisson and Thomson shared the 1937 prize "for their experimental discovery of the diffraction of electrons by crystals."[19] To make for an interesting family dinner conversation, J.J. Thomson was awarded the Nobel Prize for proving electrons are particles; his son received the award for demonstrating their wavelike properties. They were both correct.

Bohr's model of the atom and de Broglie's hypothesis of wave-particle duality were reconciled by German physicist Werner Heisenberg with his theory of quantum mechanics published in 1925, when he was only 23 years of age. Two years later, Heisenberg followed with his principle of uncertainty, which was published in the paper "*Uber den anschaulichen Inhalt der quantentheoretischen Kinematik und Mechanik*" (On the perceptual Content of quantum theoretical Kinematics and Mechanics) in 1927. The uncertainty principle states that both the position and velocity of a particle cannot be known with certainty. If our measurement of velocity is made increasingly precisely, our knowledge of the position of that same particle is correspondingly less precise. Similarly, if the position is known with precision, then the velocity must be less well known. Heisenberg derived an equation that described the minimum errors in measuring both the location and velocity of an electron. The Nobel Prize in Physics in 1932 was awarded to Heisenberg "for the creation of quantum mechanics, …".[20]

Because the energy of an electron in an atom is *precisely* known then, according to Heisenberg, we must be *uncertain* about its location. If electrons moved in simple orbits as proposed by Niels Bohr, then both their position and velocity could be determined exactly at any instance in time. According to the uncertainty principle this situation does not correspond to reality. Therefore, when discussing the motion of an electron, of known energy or velocity around a nucleus, it is possible to speak only in terms of the *probability* of finding that electron in a particular position. If all the possible positions where an electron had been, each indicated by a small dot, were summed it would create a three-dimensional cloud representing the motion of the electron around the nucleus. The outer border of the cloud is a region encompassing a probability where an electron will spend upwards of 90% of its time. This cloud represents an electron orbital.

The year 1926 was a particularly intense period of research during which several important quantum theories were proposed that accurately described the energy levels of electrons in atoms. It was during this period that Austrian physicist Erwin Schrödinger advanced his famous equation that relates the energy of an electron to its wave-like properties, which are described by a parameter called a wavefunction given the Greek symbol Ψ. The Schrödinger wave equation is rather complicated and not for the mathematically challenged, but its importance cannot be overstated. It is as central to quantum mechanics as Newton's equations are to classical mechanics.[21] The wavefunction describes the motion of an electron in an atom and is used by scientists to understand the nanoscale world. Erwin Schrödinger shared the 1933 Nobel Prize in Physics with British physicist Paul Dirac "for the discovery of new productive forms of atomic theory."[22]

While Werner Heisenberg's uncertainty principle introduced the concept of electron orbitals, Erwin Schrödinger gave them shapes. Solving the Schrödinger equation for the simplest of atoms—hydrogen with its single proton and one electron—results in a quantization permitting only certain electron orbitals and energies. These orbitals have distinct shapes and volumes, which enclose a three-dimensional space where there is a reasonably high probability (better than 90% as mentioned already) of finding the electron. But as a result of the uncertainty principle, we don't know exactly where the electron is within the orbital. The three lowest energy orbitals are designated s, p, and d. The s orbital, of which there is one, is spherical. The p orbitals, of which there are three, are shaped like weight-lifters dumbbells, while the five d orbitals, illustrated in Fig. 2.2 generally resemble four-leaf clovers.

Because these orbitals are representations of the probability of finding an electron in a particular region of space, we can perform various mathematical operations on their wavefunctions. One of the most useful, and simplest, operations we can do is to add the wavefunctions on a single atom together. This process is called hybridization, a term very familiar to plant breeders and farmers who for centuries have cross-pollinated plants to produce a hybrid product with the desired combination of characteristics.

One of the most important examples of hybridization of electron orbitals is summing s and p wavefunctions to produce sp hybrid orbitals. If we consider an atom of carbon, we can sum its outer s orbital wavefunction with the three p orbital wavefunctions to produce four sp^3 hybrid orbitals. Because of the mutual repulsive forces between electrons these four hybrid orbitals point towards the corners of a regular tetrahedron as illustrated in Fig. 2.3.

Chemists use hybridization, for instance, to explain the tetrahedral coordination of carbon atoms in diamond. The crystal structure of diamond was one of the first to be revealed using what at the time was the relatively new

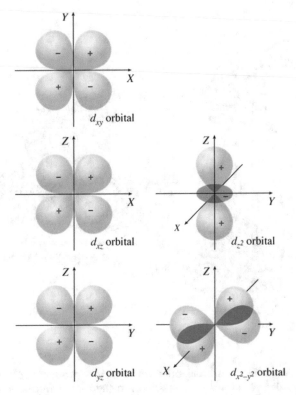

Fig. 2.2 Shapes of the five 3*d* atomic orbitals. Each three-dimensional orbital has a unique orientation as indicated by the labels on each axis. The + and − signs do not represent the charge of the electron orbital, which is always negative, but rather correspond to the sign of the wavefunction Ψ. The subscript refers to how we describe the orientation of the orbital (Reprinted from Carter CB, Norton MG (2013) Ceramic Materials: Science and Engineering, 2nd edn. Springer, New York, p 41)

technique of X-ray diffraction.[23] Using diamonds borrowed from the Mineralogical Laboratory at the University of Cambridge, Bragg father and son while both at the University of Leeds showed that the crystal structure of the precious gemstone was "extremely simple" consisting of an array of corner sharing carbon tetrahedra. But they were not able to determine the reason for the tetrahedral arrangement. Looking at the direction of the sp^3 hybrid orbitals it is easy to understand why diamond adopts the crystal structure that it does. An understanding made possible with quantum mechanics.

Considering the position of the hybrid orbitals in diamond—as far apart as possible in three dimensions—sheds light on why diamonds are so hard (they are the hardest of all known materials). The hybrid orbitals, which are each negatively charged resist attempts to push them closer than the equilibrium separation angle of 109°.

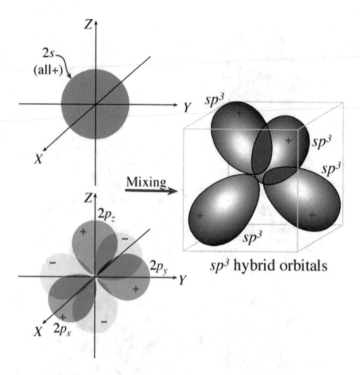

Fig. 2.3 Summing *s* and *p* orbitals on a carbon atom produces *sp³* hybrid orbitals that point towards the corners of a tetrahedron (Reprinted from Carter CB, Norton MG (2013) Ceramic Materials: Science and Engineering, 2nd edn. Springer, New York, p 63)

Graphite, another form of carbon, involves sp^2 hybridization rather than sp^3 hybridization. Where the sp^3 hybrid orbitals point towards the corners of a tetrahedron, the sp^2 hybrid orbitals point towards the corners of a triangle. The result is that graphite has a layer structure comprised of carbon hexagons with each layer connected to the ones above and below by weak van der Waals bonds. When we write with a pencil that has a graphite "lead" we leave behind a trail of carbon as we shear off layers of graphite. Graphite's layered structure, which was determined by John Desmond Bernal using X-ray diffraction, is particularly useful because we can slide atoms in between the carbon layers. For instance, in a lithium-ion battery, whether used to power a Tesla or an iPhone, graphite is the anode electrode because it is possible to place a large number of tiny lithium atoms into the gaps between the layers. The lithium ions move in and out of the graphite structure, without causing any damage, as the battery is used and then recharged.

In addition to the three quantum numbers that come from solutions to the Schrödinger equation there is an important fourth quantum number discovered by Austrian-born physicist Wolfgang Pauli in that same pivotal period from 1924 to 1927. This quantum number was needed to explain the order in which the electrons fill the various orbitals. In a many-electron atom—for instance, carbon, which has six electrons—there are multiple possible orbitals where the electrons can be placed. It is the filling of these orbitals that determines how atoms bond to each other and the properties of the resulting materials that contain those atoms.

In 1925, George Uhlenbeck and Samuel Goudsmit introduced the concept that electrons in atoms are spinning. The direction of spin—whether it is clockwise or counterclockwise—allows two electrons that have the same energy to be uniquely defined. One is spinning in one direction, whilst the other electron, with the same energy, is spinning in the opposite direction. This led to Pauli's contribution to how electrons are arranged in an atom, which was critical because it allowed an explanation of many of the important properties of materials. For instance, we would not be able to explain why iron is magnetic, which is a very important property, without using Pauli's Principle. In Chapter 8 we will see why electron spin is critical in the operation of a quantum computer. Pauli was awarded the 1945 Nobel Prize in Physics "for the discovery of the Exclusion Principle, also called the Pauli Principle."[24]

By the end of the 1920s a quantum mechanical description of atomic structure was complete. It was possible to explain why metals exhibited very high electrical and thermal conductivities. It was also possible to understand why some materials are weakly magnetic whilst others like iron and cobalt are strong, permanent magnets. The quantum mechanical model allowed us to understand why gold has the property that has marveled humans for thousands of years, its ability to resist tarnishing. While iron, the second most abundant metal on Earth, rusts when exposed to the atmosphere. Following the invention of the transistor, quantum mechanics could explain how semiconductor devices operated and help in the design of new more powerful technologies.

Embracing de Broglie's hypothesis with its implications for atomic structure based on the wavelike properties of electrons, where their specific location cannot be precisely determined, led to the development of novel electronic devices whose behavior could not be explained in terms of classical physics based on Isaac Newton's laws of motion formulated in the seventeenth century. One example is the tunnel diode that was invented by Japanese physicist Leo Esaki while at Sony Corporation in 1957 and for which he shared the Nobel Prize in Physics in 1973 for "experimental discoveries regarding tunneling phenomena in semiconductors".[25]

In a tunnel diode, electrons appear to tunnel through an otherwise prohibitive and insurmountable energy barrier rather than having to climb over it as classical physics would require. Explaining electron tunneling would not be possible without invoking their wavelike behavior. In the late 1950s Esaki's tunnel diodes offered the promise of a high-speed semiconductor switch, an important component of a computer. Unfortunately, as with a number of other device technologies developed around that time, the tunnel diode never reached its early promise. Its performance was surpassed by competing technologies, specifically the integrated circuit that was not only faster, but smaller, cheaper, and more reliable. But tunneling phenomena was important enough for Esaki to be recognized by the Swedish Academy of Sciences.

The decade of the Twenties produced some of the greatest science of any period in human history. From 1924 with de Broglie's hypothesis through to Heisenberg's uncertainty principle published in 1927 a scientific revolution took place that utterly changed our view of the atom. Many of the insights discovered and published in the Twenties were recognized by the Nobel Foundation with Albert Einstein, Niels Bohr, and Louis de Broglie all receiving the Nobel Prize in Physics during this decade. Heisenberg, Schrödinger, Davisson, and Paget Thomson received their recognition during the Thirties.

Among the important advances in materials science during the Thirties was the invention and subsequent commercialization of the transmission electron microscope by Ernst Ruska, a topic considered in more detail in Chapter 3. The use of electrons rather than light enabled an unprecedented resolution revealing material defects that had previously only been theorized. By 1970, electron microscopes were produced that were so powerful that even individual atoms could be resolved, making these instruments the imaging workhorse of nanotechnology. The Twenties marked a period of intense discovery and innovation commencing after the end of First World War. During the Thirties brand new materials would be developed that would change the world forever and create one of the major environmental challenges we face today. Plastics. And, once again there would be global conflict that would engulf the entire planet with a destructive power far exceeding that of the Great War.

Notes

1. The Nobel Prize in Physics 1929. NobelPrize.org. Nobel Media AB 2020. Sat. 12 Dec 2020 https://www.nobelprize.org/prizes/physics/1929/summary/.
2. Thomson JJ (1897) Cathode rays. Philosophical Magazine and Journal of Science 44(269).
3. The Nobel Prize in Physics 1906. NobelPrize.org. Nobel Prize Outreach AB 2021. Tue. 24 Aug 2021. https://www.nobelprize.org/prizes/physics/1906/summary/.
4. Bohr subsequently sold his Nobel medal by auction on March 12, 1940, to raise money for the Finnish Relief Fund. The medal now sits in the Danish Historical Museum of Frederiksborg.
5. There is no record of Thomson ever referring to his model as the "plum-pudding" atomic model, but the description has become mythologized and widely used both within and outside the scientific community. Hon and Goldstein give an interesting account of the term's first use, which wasn't by Thomson. Hon G, Goldstein BR (2013) J.J. Thomson's plum-pudding atomic model: The making of a scientific myth. Annalen der Physik (Berlin) 525:A129–A133.
6. da C Andrade EN (1964) Rutherford and the Nature of the Atom. Doubleday, Garden City, New York.
7. Ernest Rutherford—Facts. NobelPrize.org. Nobel Media AB 2020. Sat. 12 Dec 2020. https://www.nobelprize.org/prizes/chemistry/1908/rutherford/facts/.
8. Bruce CF, Hill RM (1960) Wavelengths of krypton 86, mercury 198, and cadmium 114. Australian Journal of Physics 14:64. The orange line is the radiation $2p^{10}$–$5d^5$. This transition cannot be explained by Bohr's model alone but can be using quantum mechanics and the orbital shapes that came from the work of Heisenberg, Schrödinger, and Pauli. The "p" orbitals are dumbbell shaped and the "d" orbitals have a clover-leaf shape.
9. Schwarz WHE (2013) 100th anniversary of Bohr's model of the atom. Angewandte Chemie International Edition 52:12228–12238.
10. Lenard PEA (1906) On cathode rays. Nobel Lecture.
11. Philipp Lenard—Biographical. NobelPrize.org. Nobel Media AB 2020. Tue. 15 Dec 2020. https://www.nobelprize.org/prizes/physics/1905/lenard/biographical/.
12. Lewis GN (1926) The conservation of photons. Nature 118:874–875.
13. Peacock G (1855) Life of Thomas Young. John Murray, London, p 148.
14. Young T (1807) A course of lectures on natural philosophy and the mechanical arts in two volumes, vol 2. Joseph Johnson, London, p 639.
15. Davisson C, Germer LH (1927) Diffraction of electrons by a crystal of nickel. Physical Review 30:705.

16. The Nobel Prize in Physics 1915. NobelPrize.org. Nobel Media AB 2020. Tue. 15 Dec 2020. https://www.nobelprize.org/prizes/physics/1915/summary/.
17. Thomson GP (1927) Experiments on the diffraction of cathode rays. Proceedings of the Royal Society A 117:600.
18. A unit cell represents the simplest repeating unit of a crystal lattice having the full symmetry of the overall crystal structure. Gold has a face-centered cubic unit cell.
19. The Nobel Prize in Physics 1937. NobelPrize.org. Nobel Prize Outreach AB 2022. Fri. 5 Aug 2022. https://www.nobelprize.org/prizes/physics/1937/summary/.
20. The Nobel Prize in Physics 1932. NobelPrize.org. Nobel Prize Outreach AB 2022. Fri. 4 Mar 2022. https://www.nobelprize.org/prizes/physics/1932/summary/.
21. Atkins PW (1978) Physical Chemistry. Oxford University Press, Oxford, p 390.
22. The Nobel Prize in Physics 1933. NobelPrize.org. Nobel Prize Outreach AB 2022. Sat. 5 Mar 2022. https://www.nobelprize.org/prizes/physics/1933/summary/.
23. Bragg WH, Bragg WL (1913) The structure of the diamond. Proceedings of the Royal Society A 89:277.
24. The Nobel Prize in Physics 1945. NobelPrize.org. Nobel Prize Outreach AB 2022. Fri. 5 Aug 2022. https://www.nobelprize.org/prizes/physics/1945/summary/.
25. Leo Esaki—Biographical. NobelPrize.org. Nobel Media AB 2020. Sun. 13 Dec 2020. https://www.nobelprize.org/prizes/physics/1973/esaki/biographical/.

3

Seeing Is Believing

In 1665, Robert Hooke, a churchman's son from Freshwater, a village at the western end of the Isle of Wight, published a book that changed our view of the natural world and caused noted diarist Samuel Pepys to call it "the most ingenious book that ever I read in my life."[1] And Pepys was nothing if not well read. What Hooke had done was use a home-made microscope to magnify tiny objects, which he then replicated by wonderfully detailed drawings published in his book *Micrographia* or to spell out its full title *Micrographia: or some physiological descriptions of minute bodies made by magnifying glasses, with observations and inquiries thereupon.*[2] At the time there was no way to photographically record the images that Hooke was seeing under his microscope. In fact, there was no way to photograph anything at all as the first photographic images would not be recorded for more than another 150 years. The only way that Hooke could document what he saw was to faithfully draw every detail.

With the aid of his microscope Robert Hooke was able to discover the microscopic structure of rocks, plants, and, most famously, insects. *Micrographia* was an archival record, available to the world, of what the microscope revealed. "A new visible World" as Hooke notes in the book's preface.

The incredible illustration of the head of a grey drone fly showed, for the very first time, the insect's amazingly complex compound eye with its array of hundreds of tiny ommatidia each able to capture and focus light. By knowing this structure, it is easy to appreciate why it is so difficult to catch a fly unawares. With its large angle view of the world, the fly is able

M. G. Norton, *A Modern History of Materials*, https://doi.org/10.1007/978-3-031-23990-8_3

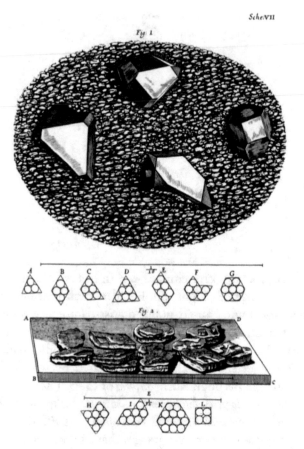

Fig. 3.1 Figures from Robert Hooke's *Micrographia*. The upper figure illustrates the faceted crystals found within broken pieces of Cornish flint. The lower figure shows the shapes of the plate-like crystals formed from dried urine (*Micrographia* is in the British Library in London, shelfmark 435.e.19)

to detect rapid movements coming from almost any direction and teasingly avoid them.

For materials scientists, one of Robert Hooke's most important observations was when he used his microscope to examine broken pieces of Cornish flint. Inside the exposed cavity Hooke found what he described as "a very pretty candied substance," which on further examination proved to be "a multitude of little *Crystaline* (*sic*) or *Adamantine* bodies, so curiously shap'd, that it afforded a not unpleasing object." Hooke's illustration of these microscopic crystals in flint is reproduced from *Micrographia* in Fig. 3.1.

The pretty crystals that so intrigued Hooke suggested that flint was a sedimentary rock comprised of a multitude of quartz crystals so small they were invisible to the naked eye. But the crystals were revealed under the lens of the microscope. As a result of its structure, when a piece of flint is broken a fracture surface remains with sharp hardwearing edges that proved such an effective tool for our earliest Stone Age ancestors. When one piece of flint is struck against another—a hammer stone against a core stone—cracks twist and turn to follow the boundaries between the densely packed quartz crystals. We describe the fracture of flint as conchoidal, or shell like, because the resulting fracture surface resembles the concave shape of a bivalve shell such as a mussel or a clam. A skilled flintknapper would remove slivers from the core until the result was a tool with a razored edge, ideal for cutting and chopping. And when attached to a wooden handle the shaped rock became an arrowhead or the tip of a spear suitable for hunting.

The regularity of the quartz crystals with their sharply defined faces led Hooke to the conclusion that they must result from a regular arrangement of particles. He was correct. These regularly arranged particles are atoms. Ancient Greek philosophers Leucippus and his pupil Democritus had described an indivisible building block of matter—the *atomos*—as early as the fifth century BCE, but it was not until 1803 that English chemist John Dalton began a one hundred and twenty three year process that would refine our understanding of what an atom is culminating in the work of Werner Heisenberg and Erwin Schrödinger in the mid 1920s that would give us our widely accepted and currently most accurate, and useful, model of the atom.

In 1849, almost two hundred years after the publication of *Micrographia*, French physicist Auguste Bravais presented his ideas on crystal structure to the French Academy of Sciences. Bravais showed that all possible three-dimensional crystals are of just fourteen distinct types.[3] Each point in these fourteen Bravais lattices represents the position of one or more atoms in the crystal. The most symmetrical of these lattices is the cube, shown in Fig. 3.2. α-polonium has a simple cubic unit cell, where a single polonium atom is placed on each corner of the cube. Gold, by contrast, has the face-centered cubic unit cell where a gold atom is located on each lattice point. Lithium is one of several metals that has the body-centered cubic unit cell. At room temperature the crystal structure of quartz is based, not on the cube, but on the hexagonal Bravais lattice with a silicon-oxygen tetrahedron sitting on each lattice point. The outer electron orbitals in silicon are sp^3 hybridized, they point to the corners of a tetrahedron, just like the hybrid orbitals in the atoms of carbon that make up diamond. If you refer back to Fig. 2.2 there is an illustration of the orientation of the sp^3 hybrid orbitals.

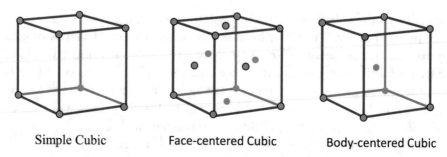

Simple Cubic   Face-centered Cubic   Body-centered Cubic

Fig. 3.2 Unit cells of the three cubic Bravais Lattices. The crystal structure is built up by stacking these unit cells in three dimensions so that they fill space. Each sphere in the illustration is a lattice point that can accommodate one or more atoms

Hooke's images of quartz are remarkable for what they reveal, but his microscope would not have been able to resolve the individual atoms that comprise those crystals.

Hooke's fascination with crystals extended to even looking at those formed from dried urine. The plate-like crystals revealed by the microscope were a mixture of shapes: rhomboids, squares, and rectangles (Fig. 3.1 lower). They also were in a range of colors "some White, some Yellow, some Red, others of more brown and duskie colours (*sic*)." Many of the crystals were faceted just like those found in the cavities of broken pieces of flint, implying that certain crystal planes—or faces—are formed preferentially over others: planes with the lowest energy being larger, bound by smaller higher-energy edges. Studying these crystals led Hooke to propose a procedure to deal with painful and troubling kidney stones. Three hundred and fifty years after *Micrographia*, images of kidney stones obtained by a team at the University of Chicago using a powerful state-of-the-art scanning electron microscope show that Hooke's observations were perfectly correct in identifying the shape of the stones.[4]

Hooke's microscope used three glass lenses, making it a compound microscope, which he described in *Micrographia* as: "The Microscope, which for the most part I made use of, … was contriv'd with three Glasses; a small Object Glass a thinner Eye Glass, and a very deep one: This I made use of only when I had occasion to see much of an Object at once; the middle Glass conveying a very great company of radiating Pencils, which would go another way, and throwing them upon the deep Eye Glass. But when ever I had occasion to examine the small parts of a Body more accurately, I took out the middle Glass, and only made use of one Eye Glass with the Object Glass." Hooke's accomplishments would have been groundbreaking even with the most perfect of modern-day glass lenses. But the lenses he used, although state-of-the-art, were far from perfect and had significant defects, which created distorted images. As Hooke himself noted in the above quote, to get more accurate observations he actually *removed* a lens to reduce the total distortion of the image. The lower magnification, perhaps counterintuitively, produced a sharper image.

To place Hooke's glass lenses within the context of glass manufacturing in 1665, Venetian glass makers working on the tiny island of Murano were producing the highest quality glass, which they turned into elaborate vases and delicate tableware. Other European manufacturers including those in Paris were pouring glass for mirrors. Shaping was usually done by casting the molten glass onto a metal table or blowing the glass into a large sphere, pressing it then spinning it to produce a flat disk, which could be cut to the desired size once cold. The beautiful Hall of Mirrors, the highlight of the Palace at Versailles, was completed in 1684 using glass panels made with the latter technique. These large glass panels much like the smaller glass lenses used for the early microscopes were a mixed alkali composition containing many defects including regions where the density of the glass was not uniform. Nonuniformity introduces local stresses, which in turn cause distortion of the image.[5]

Optical glasses of the compositions used today for lenses, prisms, and other applications relying on the transmission of light were not available to Robert Hooke when he was examining the specimens that would be recorded in *Micrographia*. The two most common types of optical glass are flint glass and crown glass. Flint glass, often called "lead crystal" even though it is not crystalline, was developed by George Ravenscroft in 1675. It was lead crystal glass that made England the leading glass producing country not just in Europe but in the entire world. Crown glass has a formulation very similar to the glass used at Versailles, but with higher amounts of potassium oxide (called potash in the glass industry.) This glass was not available in London until 1678, more than a decade after the publication of *Micrographia*. Flint glass and crown glass have complementary properties. When cemented together they produce a lens called an achromatic doublet, with a very low chromatic aberration.[6] The achromatic doublet became widely used in the late 1750s for both microscopes and telescopes. Hooke died in 1703, before he would have been able to take advantage of these innovations in optics.

Minimizing or compensating for lens aberrations whether in the glass lenses used for light microscopy or in the electromagnetic lenses used in electron microscopes represents the most important method for increasing our ability to resolve smaller and smaller features. In electron microscopy, electromagnetic lenses have improved to such a point that atomic resolution is almost routine.

The first images of individual atoms would not be obtained for more than 300 years after the publication of *Micrographia*. These images would be of two uranium atoms, 0.46 nm in diameter, attached to either end of an organic molecule. In 1970 British-born American scientist Albert Crewe, a professor of physics at the University of Chicago, was able to record these atomic images using a powerful electron microscope.[7] To the untrained eye the images are unimpressive—fuzzy regions of light and dark grey-scale contrast. But as Crewe reflected upon the impact of his achievement: "The ability to see such atoms may make it possible to determine shapes of molecules and their relationships. In addition, the chemistry of these techniques could be seen and the precision of atomic locations and the degree of reaction could be studied." Crewe added that the visibility of atoms and their arrangement in molecules should enhance greatly many fields, especially medicine, biochemistry, and genetics. The technique should be particularly valuable in analyzing chromosomes and cancer cells.[8] Crewe was right!

The term 'resolution' describes the ability of a microscope to distinguish fine detail. Our eyes with good lighting can typically resolve features separated by about 0.1 mm. If two objects for instance points on a piece of paper or pixels on a computer screen were only half that distance, 0.05 mm apart, the human eye would in most cases not be able to differentiate them. They would appear as a single point. A microscope has a higher resolution than the human eye. In other words, using a microscope we can see—or resolve—features beyond that of the unaided eye. But even with the highest quality lenses and the brightest possible illumination we cannot resolve individual atoms using light. The ultimate resolution of a light microscope is limited by the wavelength of light and by diffraction effects that cause a point of light to appear not as a single sharply defined point, but rather as a central bright disc surround by a series of concentric rings. These rings—the result of diffraction—are named after their discoverer English mathematician and Astronomer Royal George Biddell Airy. Although Airy was using a telescope for distant observations of stars rather than looking at the minute through a microscope, he was still imaging light using glass lenses.

In 1835, Airy published his observations in the *Transactions of the Cambridge Philosophical Society* in a paper entitled "On the diffraction of an object-glass with circular aperture" in which he described the rings that he saw surrounding the image of a star: "The image of a star will not be a point but a bright circle surrounded by a series of bright rings."[9] Over 80%

of the light is contained within the central disc with the remaining 20% in the diffraction pattern. Every point of light coming from an object—whether it is near or far—is diffracted into a pattern of Airy rings. It is the overlap of these patterns that introduces a fundamental resolution limit for all optical instruments. The disk pattern observed by Airy closely resembles the electron diffraction pattern shown earlier in Fig. 2.1, they are both the result of diffraction, a property of waves.

In 1873 German physicist Ernst Karl Abbe published his theory and corresponding mathematical formula defining the diffraction limit.[10] Abbe's equation stated that the ultimate resolution of a light microscope was approximately equal to one half that of the wavelength of the imaging light source. Visible light wavelengths range from the very short violet at 380 nm all the way to red at a wavelength of 700 nm. White light is a mixture of all visible wavelengths from violet to red. To maximize the resolution of a light microscope the specimen should be viewed with light of the shortest possible wavelength. So, in principle, the highest resolution in a light microscope using only violet light illumination is limited to around 200 nm. This is five hundred times better than the unaided human eye, but not sufficient to resolve the atoms that make up the Cornish flint crystals that Hooke was carefully examining.

When referring to the resolution limit of their instruments microscopists often use the "Rayleigh criterion" invented by John William Strutt (3rd Baron Rayleigh), which defines the smallest resolvable feature in an imaging system limited by diffraction. The Rayleigh criterion states that if the center of the central Airy disc of one point of light overlaps the first minimum of the diffraction pattern from an adjacent point, then we have reached the resolution limit. The two points are then just resolvable. If the points are any closer than the Rayleigh criterion permits they cannot be resolved, appearing as just a single point.

Realizing that there was a physical limit for the resolution of a microscope must have been a great disappointment to microscopists at the time of Abbe's work. This disappointment was perhaps felt particularly strongly by Abbe himself who complained that "it is poor comfort to hope that human ingenuity will find ways and means of overcoming this limit."[11]

While the earliest microscopes had been built using the trial-and-error method with homemade lenses, from 1872 on, shortly after Abbe was appointed as extraordinary professor of physics and mathematics at the University of Jena, they were designed on the foundation of sound scientific calculations together with optimized optical glass formulations. These microscopes with their vastly improved optical properties enabled the pioneering research of the late nineteenth century in several fields particularly biology and medicine. Two examples are the work of Robert Koch and Paul Ehrlich. An important aspect of imaging biological systems such as Koch's work with bacteria and Ehrlich's work with antibodies is the need to "stain" the sample to make it easier to see some of the finer microstructural features. One way to stain biological samples is to apply a dye such as crystal violet. The dye binds to proteins and DNA and can be used, for instance, to measure cell death in a culture caused by chemicals, drugs, or toxins from pathogens.

Similar contrast-enhancement techniques needed to be developed to study materials—mainly metals—using a light microscope. In most cases the structure of a metal would be revealed by immersing it in an etching reagent. This attacks the grain boundaries that separate one grain from another allowing individual grains to be distinguished, their size, shape, and orientation determined. Nitric acid, a powerful acid and oxidizing agent known as "aqua fortis" (strong water) in Latin, is an excellent etchant for iron and steel alloys allowing the different phases—pearlite, ferrite, and martensite—to be clearly identified.[12] A major difference between looking at biological specimens and studying metals is that biological microscopes use transmitted light, while metallurgical microscopes, like Hooke's compound microscope, on the other hand must use reflected light. Unless they are extremely thin, metals are opaque to light.

Since the publication of *Micrographia* the use of light microscopy has led to many advances in medicine, the life sciences, metallurgy, and the physical sciences. However, a persistent limitation has been its resolution. According to Abbe's formula to improve the resolution of a microscope beyond that possible using visible light a form of illumination was needed that had a shorter wavelength.

Enter the electron.

As mentioned earlier in Chapter 2, that electrons are waves had been hypothesized by Prince Louis-Victor Pierre Raymond de Broglie in 1924. Experimentally, the wavelike properties of electrons were demonstrated three years later by Bell Labs researchers Clinton Davisson and Lester Germer who showed that an accelerated beam of electrons could be diffracted (a defining property of all waves) by a crystal of nickel. The wavelength of the electron beam used by Davisson and Germer in their classic study was about 0.1 nm[13]—more than four hundred times shorter than that of visible light.

Electron wavelengths become compressed as they are accelerated using higher and higher voltages. In a modern transmission electron microscope operated at around 300,000 V, the electrons are moving at a very significant fraction of the speed of light and have a wavelength of only 1.96 pm

(0.00196 nm, much smaller than the average size of an atom). The wavelength of a beam of electrons is so much shorter than that of visible light, which makes the theoretical resolution of an electron microscope essentially unlimited in its ability to image atomic-level structure. In practice, the resolution of an electron microscope is limited because it is impossible to make perfectly uniform electromagnetic lenses that are necessary for precisely focusing an electron beam. Even with this limitation, the best electron microscopes can resolve features smaller than 0.05 nm (½ Ångstrom).[14]

The invention of the electron microscope by Max Knoll and his student Ernst Ruska in 1931 finally overcame the barrier to better resolution that had been imposed by the limitations of visible light and had so disappointed Ernst Abbe.[15] Ruska's pioneering work on the electron microscope began in the 1920s while he was a young student at the Technical College in Berlin. There he found that a magnetic coil could be used to focus electrons, in a similar manner to how glass lenses focus light. By coupling two electromagnetic lens, Ruska produced a primitive electron microscope. With this rudimentary, but nonetheless groundbreaking, instrument Ruska demonstrated that he could obtain an image of an object, a molybdenum metal mesh, irradiated by electrons. A photographic plate placed below the sample captured the image. The magnification possible with this first microscope was only a modest seventeen times, certainly within the range of a good light microscope. By 1933 Ruska was able to improve on the design sufficiently to build *by himself* an electron microscope with a performance clearly superior to that of the conventional light microscope. And by 1938 Ruska had produced a prototype of a commercial instrument with a magnification of 30,000 times. One year later the first commercially available mass-produced electron microscope, the "Siemens Super Microscope", entered the market.[16] This microscope had a resolution of about 10 nm. By the mid 1940s this was further reduced to just 2 nm—several orders of magnitude better than the theoretical resolution of a light microscope.

Ruska's electron microscope is a *transmission* electron microscope or TEM. The object to be examined is thin enough that it is transparent to the beam of electrons. Because of how strongly electrons interact with matter, to be electron transparent any specimen to be examined must be very thin. Ideally a specimen for TEM studies should be less than 100 nm, although 10 nm or thinner is even better, particularly if the goal is to achieve atomic resolution.

With the earliest electron microscopes it was not possible to see individual atoms—the ultimate goal—but they were able to provide images of materials with a resolution beyond what had previously been possible, or even imagined, at the beginning of the century. With a resolution of about 10 nm these

Fig. 3.3 Magnesium oxide "smoke" crystals made by burning a strip of magnesium metal. The cubic shape of the crystals is clearly seen in the image (Reprinted from Pikhitsa PV, Chae S, Shin S, Choi M (2017) A low-field temperature dependent EPR signal in terraced MgO:Mn^{2+}nanoparticles: an enhanced Zeeman splitting in the wide-bandgap oxide. Journal of Spectroscopy 8276520 with permission)

early instruments were able to be used to study small particles and determine, for instance, their size, shape, and distribution.

One area of materials science where early electron microscopes were able to provide important new insights and understanding was in colloidal systems. Colloid is a term encompassing a very wide range of materials. It is generally applied to describe any two-phase substance where one of the phases consists of tiny particles—typically in the range of 5 to 1,000 nm—dispersed through a second substance. Perhaps, an easily visualized example of a colloidal system is smoke particles in air. Particles in wood smoke are on average about 500 nm across. As an additional example, "smoke" particles of magnesium oxide formed by burning magnesium metal in air are typically only 100 nm or even less in diameter. This is much smaller than the range of even the best light microscope. The magnesium oxide particles appear as shown in Fig. 3.3 as perfect cubes in the electron microscope. Magnesium burns a very bright white and is used as a common constituent of fireworks to add white sparks and improve the overall brilliance.

Rubber is a commercially important colloidal system. It is essential in many applications and was labelled a "strategic and critical material" by President Franklin Roosevelt at the beginning of the Second World War.[17] None of the applications for rubber, in war or peacetime, exceed its importance for tires. In 1912, carbon black was added to natural rubber because it was found to increase the strength of automobile tires. The exact mechanism that produced this strengthening was a mystery. At the time there were no techniques that could "see" the minute carbon black particles and how they were dispersed throughout the material. However, when viewed in the electron microscope the carbon black particles could be seen as smooth spheres varying in size between 10 and 400 nm in diameter.[18] The size distribution of the carbon black particles was found to be an essential factor in determining the mechanical properties of the rubber. Tires with small carbon black particles were more hard-wearing than those containing larger particles. The strength of the rubber increased uniformly as the dispersed carbon particles got smaller. In addition to improving its mechanical properties, the presence of carbon black increased the electrical conductivity of the rubber reducing the build-up of static electricity. Fillers, including carbon black with an average particle size of about 35 nm, make up over 25% of the formulation of modern automobile tires.

Another colloidal system whose structure was revealed by electron microscopy was the origin of the bright ruby red color known as the "Purple of Cassius", which was used as early as Roman times to produce a red glass. The color was found to be due the presence in the glass of approximately spherical gold particles with an average diameter of 20 nm. These microscopic features would certainly not have been known to the Romans, but that did not prevent them from making striking examples of ruby-colored glasses and even using mixtures of gold and silver colloids. The fourth century Lycurgus cup, which is on display in the British Museum in London, illustrates the story of Lycurgus the ill-fated king of the Thracian Edoni who was strangled by vines after taunting the god Dionysus. The color of the glass is particularly unusual because it appears deep ruby red when viewed in transmitted light, as shown in Fig. 3.4. In reflected light the cup is pea green. This effect known as dichromatism (two colors) is the result of both gold and silver colloids being present within the glass.[19]

One person who made significant contributions to the scientific understanding of colloids was Austrian scientist Richard Zsigmondy. Zsigmondy was born in Vienna in 1865 and spent most of his career as a professor at the University of Göttingen in Germany. In 1925 Zsigmondy was awarded the Nobel Prize in Chemistry "for his demonstration of the heterogeneous nature of colloid solutions and for the methods he used, which have since become fundamental in modern colloid chemistry."[20] Among the materials Zsigmondy studied was the color produced by gold colloids, the purple of

Fig. 3.4 A beautiful example of a colloidal system, the Lycurgus drinking cup (late Roman) housed in the British Museum in London. In this picture the cup is viewed in transmitted light. The late Roman Lycurgus drinking cup is housed in the British Museum in London. (Reproduced with permission)

Cassius, which he described in his book *Das Kolloide Gold*, a collaboration with P.A. Thiessen published in the same year the Nobel Prize was awarded.[21]

A significant observation Zsigmondy made was that as the size of the gold particles increased there was a noticeable change in color from red to blue. This result supported the earlier observations of English scientist Michael Faraday in the mid 1800s. Richard Zsigmondy reflected during this Nobel Lecture delivered in December 1926 that he was at the time unaware of Faraday's results commenting: "If I had known of Faraday's results, it would have saved me much unnecessary work."

Among the specific questions Zsigmondy was attempting to answer were: What are the sizes of the gold particles? Are they molecules or aggregates

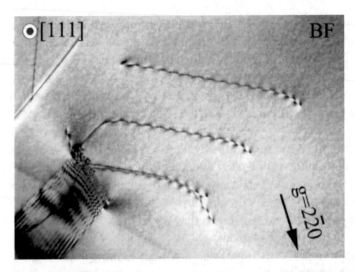

Fig. 3.5 High voltage transmission electron microscope image of dislocations being emitted ahead of a crack tip in a crystal of silicon. The numbers indicate the orientation of the sample and the electron beam. The label BF stands for bright field. In a dark field (DF) image the contrast would be reversed. (Reprinted from Tanaka M, Higashida K (2004) HVEM characterization of crack tip dislocations in silicon crystals. Journal of Electron Microscopy 53:353 with permission from Oxford University Press)

of molecules? How many atoms approximately does a single gold particle contain? Using the "ultramicroscope" that Zsigmondy developed he was able to see gold particles with diameters as small as 10 μm, but this is far larger than those that have been identified in the Roman red glasses using electron microscopy.

Another area of materials science where the transmission electron microscope has been indispensable, and without equal, is in identifying and studying dislocations. Dislocations are line defects, rows of displaced atoms, within the internal structure of a material. Their importance in metals is that plastic deformation, a permanent change in shape for instance when a metal wire or sheet is bent, occurs by the creation and movement of a large number of dislocations. Dislocation movement in a process called slip is why metals are weaker than theory would predict. Ceramics, on the other hand, which are frequently brittle materials, are so because they generally have very few dislocations. Those that are present are often extremely difficult to move.[22] Figure 3.5 is a transmission electron microscope image of dislocations in a crystal of silicon. The dislocations appear as jagged dark lines, resembling a row of medical stitches, being emitted from the tip of a crack. Many more dislocations would be seen in a typical transmission electron microscope image of a metal.

The concept of a dislocation was first introduced by Charles Burton in 1892 as a "modification in the structure … of the aether, and when matter moves it is merely these modifications of structure … which are transferred from one portion of the aether to another."[23] The term "dislocation" referring to an atomic-scale defect was coined by Cambridge mathematician and physicist Geoffrey Ingram Taylor in 1934 to describe the defect created by slip.[24]

Dislocations were not only invoked to explain how metals plastically deform they could also be used to explain the effect of mechanical processes such as work hardening. During work hardening, for instance by repeatedly bending a piece of wire, so many dislocations are introduced into the metal that they get in each other's way as they try to move. This causes them to pile up and resist further movement.

Dislocations are also important in understanding how crystals grow. A dislocation terminating on the surface of a material can create an exposed "cliff'" of atoms that acts as a nucleation site for further growth.[25] The importance of dislocations in crystal growth and in determining the mechanical properties of materials stimulated the search for techniques to observe the processes directly inside the electron microscope.

Using the new, made in 1954, Siemens Elmiskop I electron microscope James Menter while working at Tube Investments Research Laboratories near Cambridge produced the first-ever image of a dislocation. It was of an edge dislocation in a very thin crystal of the compound platinum phthalocyanine. Although poorly resolved by today's standards Menter was able to identify planes of atoms magnified 1,500,000 times that were spaced only 1.2 nm apart.[26] The image was recorded in December 1955 and published the following year in the *Proceedings of the Royal Society of London*. Direct imaging of the dislocation was possible in platinum phthalocyanine because the building block, the unit cell, of the crystal structure is relatively large. The Siemens Elmiskop I would not have had the resolution to directly observe edge dislocations in more simple crystals, with smaller unit cells, such as aluminum and gold. In these close-packed cubic structures the atomic planes are much more closely spaced than they are in the metal phthalocyanine compounds. A different approach was needed to observe dislocations in close-packed metals.

With a Siemens Elmiskop I, similar to Menter's, at the Cavendish Laboratory in Cambridge Peter Hirsch and his doctoral student Michael Whelan revealed dislocations in an aluminum foil utilizing a technique in the electron microscope called diffraction contrast where the dislocations appear as dark lines often forming symmetric grid patterns or arrays.[27] As with the image in Fig. 3.5 the dislocations are revealed in stark contrast to the lighter dislocation-free background. Because it is possible to tilt the sample inside an electron microscope the exact orientation of the dislocations can be determined. With this information we know on what planes the dislocations slip and the direction of that slip.

What was equally important to Hirsch and Whelan's static observation of dislocations was that in the microscope it was actually possible to see how the defects moved. Working at high electron beam currents to heat the aluminum sample two types of dislocation motion were observed: rapid and slow and jerky. This was a first. No one had ever watched dislocations slip. These observations helped to significantly expand our understanding of the mechanical properties of metals, properties that are dominated by the movement of dislocations.

By the 1970s with significant advances in the quality of the electromagnetic lens and improvements in housing microscopes to reduce mechanical vibrations, a point resolution of only 0.3 nm was achieved. This is on the scale of the spacing between planes of atoms in a close-packed metal. For instance, the close-packed planes in aluminum are separated by 0.23 nm. A major step forward in direct imaging of the atomic structure in solids was the construction of the Atomic Resolution Microscope (ARM) that was put into operation in 1983 at the Lawrence Berkeley Laboratory in California. Towering over two stories tall this microscope could operate at electron accelerating voltages up to 1 million volts enabling a sub-nanometer resolving power better than 0.16 nm. This level of performance made the ARM the first instrument in the world able to distinguish individual atoms in a solid.

A major concern with locating such a delicate instrument in northern California was the distinct possibility of earthquakes. Berkeley lies very close to the San Andreas fault, one of the largest geological faults in the world. To stabilize the 30-ton giant of a microscope it was embedded in one hundred tons of cement, which lowered its center of gravity. It was floated on air springs and tethered to what is described as a sort of combination pogo-stick and disk-brake foundation. These elaborate precautions were designed to keep the microscope from walking away in up to a 7.8-magnitude quake.[28]

In 1984, the National Center for Electron Microscopy (NCEM) was established at the Lawrence Berkeley Laboratory to provide a shared resource to researchers from around the world. At the heart of NCEM was the atomic resolution microscope together with a high-voltage electron microscope that operated at 1.5 million volts, making it one of the most powerful microscopes ever made.

Since their invention in the 1930s, electron microscopes have helped scientists peer into the atomic structure of ordinary, but very complex, materials including steel, concrete, and rubber. They have opened an entire world that exists on the nanoscale from gold nanoparticles to exotic new materials such as graphene. Notwithstanding these incredible advances, two challenges remained: how to obtain three-dimensional images of atomic structure and how to watch materials behavior in the microscope under real-life operating conditions. Recently, a multinational collaboration including researchers at the National Center for Electron Microscopy at Lawrence Berkeley Laboratory, the Institute for Basic Science in South Korea, Monash University in Australia, and the University of California Berkeley have developed a technique that produces atomic-scale three-dimensional images of platinum nanoparticles tumbling in liquid held between sheets of graphene, the thinnest material possible at just one atom thick.[29] Each platinum nanoparticle is less than 3 nm in diameter, comprising approximately 600 atoms. A static image of one of these nanoparticles in three different orientations is shown in Fig. 3.6. Next to the images are the reconstructed atomic position maps, which have a resolution better than 0.072 nm. From these images it is easy to discern the unit cell shapes. Scientists are excited about this research because the properties of nanomaterials depend on the 3D atomic arrangement. Even small deviations in structure, particularly at or near the surface can have a significant impact on the properties, and usefulness, of nanomaterials. That disorder is most clearly seen on the periphery of the particles shown in the three images on the left-hand side of the figure.

Ernst Ruska was awarded a half share of the 1986 Nobel Prize in Physics for "his fundamental work in electron optics, and for the design of the first electron microscope."[30] For many of us who have used electron microscopes in our research and to Ruska himself the fifty-five year wait for recognition seems excessive particularly considering that the Royal Swedish Academy of Sciences in the press release announcing the prize stated that the electron microscope "is one of the most important inventions of this [20th] century." Without the electron microscope the entire field of nanotechnology would be entirely empirical lacking the in-depth scientific understanding that high resolution electron microscopy has enabled. Ruska died in 1988 only two years after receiving the award, which is never given posthumously.

Sharing the other half of the 1986 Nobel Prize in Physics with Ernst Ruska were Gerd Binnig and Heinrich Rohrer for their "design of the scanning tunneling microscope." At the time of their groundbreaking research, Binnig and Rohrer were both at the IBM Zürich Research Laboratory in Rüschlikon, Switzerland. The "STM" as it is widely known is a very different type

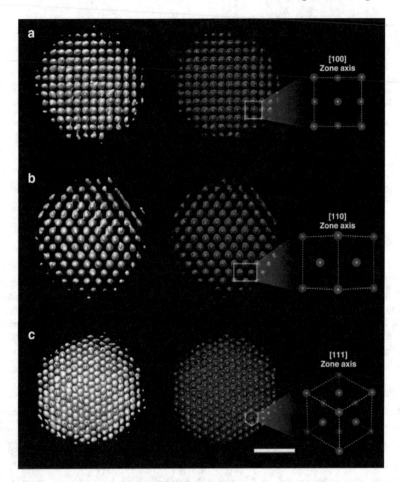

Fig. 3.6 The ultimate in atomic resolution; images of platinum nanoparticles. The scale bar on the bottom right is 1 nm long. At this resolution the unit cell shapes can be seen from the images, which also show that closer to the surface the atomic arrangements are increasingly disordered (Reprinted from Kim BH, Heo J, Kim S, Reboul CF, Chun H et al. (2020) Critical differences in 3D atomic structure of individual ligand-protected nanocrystals in solution. Science 368:60-67, with permission The American Association for the Advancement of Science)

of microscope from the electron microscope invented by Ruska. In fact, it is not a true microscope in the traditional sense—an instrument that gives a direct image of an object. But what the two techniques do share is the ability to achieve atomic resolution, to image individual atoms.

How a scanning tunneling microscope works has more in common with vinyl record players than with focusing and imaging an electron beam in a transmission electron microscope. In a scanning tunneling microscope an

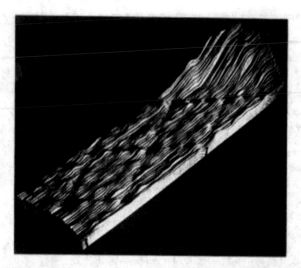

Fig. 3.7 Scanning tunneling microscope image of the surface of a silicon crystal. This image provided experimental proof for the rearrangement of the silicon atoms on the surface of the material. A voltage of 2.9 V was maintained between the metal tip and the silicon surface to avoid them touching. (Reprinted with permission from Binnig G, Rohrer H, Gerber Ch, Weibel E (1983) Physical Review Letters 50:120. Copyright 1983 by the American Physical Society)

extremely sharp tungsten stylus scans the surface of the material to be examined without ever touching it. The tip of the stylus floats above the sample surface at a distance of about 1 nm. To maintain this separation, a constant electrical current is maintained between the stylus and the surface of the material. This current is due to electrons tunneling across the nanometer-sized gap. As mentioned in Chapter 2 tunneling is a quantum mechanical phenomenon where electrons can exist on both sides of a very thin electrically insulating barrier, a situation that could not be explained by classical physics. By precisely recording the vertical movement of the stylus as it maintains a constant tunneling current while scanning the surface of the sample it is possible to study surface structure atom by atom.

The first scanning tunneling microscope image obtained by Binnig and Rohrer was of the surface of silicon, shown in Fig. 3.7.[31]

The silicon atoms can be clearly resolved by the contours, like lines on a map, which rise to form an array of small bumps. What particularly excited surface scientists about this image was that it provided direct experimental evidence of the reconstruction of the silicon surface.[32] An explanation for this reconstruction, one of the most intriguing and complex problems in the field of surface science, had eluded many theoretical approaches. As the group in Zürich described: "In order to make significant progress, some basically

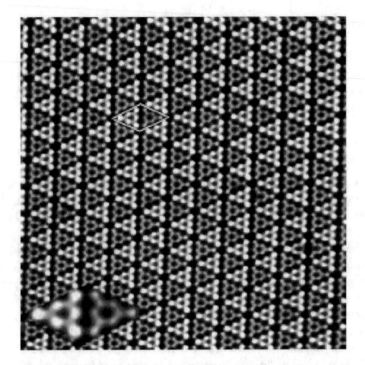

Fig. 3.8 Scanning tunneling microscope image of the silicon surface recorded at the Chinese Academy of Sciences in Beijing. The reconstruction of the surface is the same as that for the sample studied by Binnig, Rohrer, Geber, and Weibel twenty-five years earlier. The enhanced resolution of the atomic structure of the surface and the use of color enhancing illustrates how the technique has advanced (Reprinted from Wang Y-L, Guo H-M, Qin Z-H, Ma H-F, Gao H-J (2008) Toward a detailed understanding of Si(111)- surface and adsorbed Ge nanostructures: fabrications, structures, and calculations. Journal of Nanomaterials 874213 with permission.)

new approach is required. Such a new approach is the *scanning tunneling microscopy* recently introduced by the authors."

Although perhaps difficult to discern, the image shows two complete surface unit cells each containing twelve atoms. A more recent image in Fig. 3.8 clearly shows the power of the scanning tunneling microscope to achieve atomic scale resolution of the silicon surface. The rhombus indicated in the high-resolution image in the bottom left is the same surface unit cell shown in Fig. 3.7.

Fifty years separate Ruska's construction of the first modern electron microscope, an instrument with considerably better performance than a conventional light microscope, and Binnig and Rohrer's landmark publication showing the atomic resolution of the reconstructed silicon surface. Both instruments have become essential tools for the materials scientist and can be found in almost every university and industry laboratory.

This chapter was about how we characterize materials. The step before we characterize a material is to make it. The next chapter describes how we make some of the materials that have shaped our modern world.

Notes

1. Samuel Pepys diary entry on Saturday 21 January 1664/65. After having his supper Pepys stayed up until two o'clock in the morning reading Hooke's book. Entry available at https://www.pepysdiary.com/diary/1665/01/.

2. Hooke R (1665) Micrographia: or some physiological descriptions of minute bodies made by magnifying glasses, with observations and inquiries thereupon, London. The first important work on microscopy. Not only does it provide detailed observations of the microscopic, but it also provides a glimpse of some of the challenges Hooke faced working with his subjects. For example, the ant (p. 203) proved to be particularly problematic and Hooke had to sedate him with 'Brandy,' which 'knock'd him down dead drunk, so that he became moveless.' The ant recovered an hour later and 'ran away.' Though Hooke did not engrave the images in *Micrographia* himself they were engraved after his illustrations.

3. Bravais A (1850) Mémoire sur les systèmes formés par des points distribués regulièrement sur un plan ou dans l'espace. Journal de L'Ecole Polytechnique 19:1–128. The fourteen Bravais lattices are derived from the seven crystal systems—Triclinic, Monoclinic, Orthorhombic, Tetragonal, Rhombohedral, Hexagonal, and Cubic. Cubic is the most symmetric and there are three cubic Bravais lattices—simple cubic, body-centered cubic, and face-centered cubic.

4. Arvans D, Jung Y-C, Antonopoulos D, Koval J, Granja I, Bashir M, Karrar E, Roy-Chowdhury J, Musch M, Asplin J, Chang E, Hassan H (2017) *Oxalobacter formigenes*-derived bioactive factors stimulate oxalate transport by intestinal epithelial cells. Journal of the American Society of Nephrology 28:876–887.

5. Window glass and that used for mirrors and other applications during the seventeenth century prior to the widespread use of crown and flint glass was a mixed alkali glass contained (in weight %)7–13 Na_2O; 3.5–7 K_2O, 8–12 CaO; 2–3.5 MgO. The remainder being primarily SiO_2. A great reference is Velde B (2013) Seventeenth-Century Varec Glass from the Great Hall of Mirrors at Versailles in: Modern Methods for Analysing Archeological and Historical Glass, 1st ed. Janssens KHA (ed) Wiley, New York p. 569. Crown glass is also a mixed alkali glass, but contains 10 wt% K_2O.

6. Chromatic aberration is where different wavelengths of visible light are not focused to the same point. The result is a color distortion or colored fringe

around the edge of objects in an image. Chromatic aberration makes the resolution of a microscope worse.

7. Crewe AV, Wall J, Langmore J (1970) Visibility of single atoms. Science 168:1338. Joseph Wall and John Langmore were graduate students in the Department of Biophysics working closely with Crewe.

8. Fling KJ (1970) First photographs of a single atom. Bio Science 20:918.

9. Airy GB (1835) On the diffraction of an object-glass with circular aperture. Transactions of the Cambridge Philosophical Society 5:283.

10. Abbe E (1973) Beiträge zur theorie des mikroskops und der mikroskopischen wahrnehmung. Archiv für Mikroskopische Anatomie 9:413–468.

11. Quoted in Williams DB, Carter CB (2009) Transmission Electron Microscopy. Springer, New York p. 5.

12. In 1863 English microscopist and geologist Henry Clifton Sorby developed etching techniques using acids for studying the microstructure of iron and steel, which led to an understanding of the role of carbon on the mechanical strength of steel. Sorby HC (1886) On some remarkable properties of the characteristic constituent of steel. Proceedings of the Yorkshire Geological and Polytechnic Society 9:145–146. Sorby's studies also paved the way for mass production of steel.

13. Davisson C, Germer LH (1927). Diffraction of electrons by a crystal of nickel. Physical Review 30:705. As mentioned in Chapter 2 G.P. Thomson and A. Reid also proved the wavelike properties of electrons. They used an electron beam that had a wavelength of only 0.01 nm—an order of magnitude shorter than that used by Davisson and Germer.

14. The Ångstrom despite not being part of the SI system of units is widely used in the field of electron microscopy and in many branches of physics to express wavelengths and interatomic distances. The unit is named after Swedish physicist Anders Jonas Ångström. 1 Å = 0.1 nm.

15. Knoll M, Ruska E (1932) Das elektronenmikroskop. Zeitschrift für Physik 78:339.

16. Ruska E (1940) Electronic Microscope. United States Patent 2,272,353.

17. The National World War II Museum, New Orleans http://enroll.nationalww2museum.org/learn/education/for-students/ww2-history/at-a-glance/rubber.html. Accessed 8 August 2022.

18. Turkevich J, Hillier J (1949) Electron microscopy of colloidal systems. Analytical Chemistry 21:475.

19. Freestone I, Meeks N, Sax M, Higgitt C (2007) The Lycurgus Cup—A Roman nanotechnology. Gold Bulletin 40:270–277.

20. The Nobel Prize in Chemistry 1925. NobelPrize.org. Nobel Prize Outreach AB 2022. Tue. 7 Jun 2022. https://www.nobelprize.org/prizes/chemistry/1925/summary/.

21. Zsigmondy R, Thiessen PA (1925) Das Kolloide Gold. Akademische Verlagsgesellschaft, Leipzig.

22. William Dash a crystal grower at General Electric Research Laboratories in Schenectady, New York, developed a method to grow silicon crystals completely free of dislocations. These crystals are used in the manufacture of silicon chips.

23. Burton CV (1892) A theory concerning the constitution of matter. Philosophical Magazine 33:191.

24. Taylor GI (1934) The mechanism of plastic deformation of crystals. Part I.—theoretical. Proceedings of the Royal Society London A 145:362–387.

25. Burton WK, Cabrera N, Frank FC (1949) Role of dislocations in crystal growth. Nature 163:398–399.

26. Menter JW (1956) The direct study by electron microscopy of crystal lattices and their imperfections. Proceedings of the Royal Society London A 236:119–135. The Siemens Elmiskop was operating with an electron accelerating voltage of 80 kV, which gave an electron wavelength of 0.00417 nm (0.0417 Å).

27. Hirsch PB, Horne RW, Whelan MJ (1956) Direct observations of the arrangement and motion of dislocations in aluminium. Philosophical Magazine 1:677–684.

28. Paul Preuss, Twenty years and still growing (smaller). Science Beat available online https://www2.lbl.gov/Science-Articles/Archive/sb-NCEM-20.html.

29. Kim BH, Heo J, Kim S, Reboul CF, Chun H, et al. (2020) Critical differences in 3D atomic structure of individual ligand-protected nanocrystals in solution. Science 368:60–67.

30. The Nobel Prize in Physics 1986. NobelPrize.org. Nobel Prize Outreach AB 2021. Tue. 31 Aug 2021. https://www.nobelprize.org/prizes/physics/1986/summary/.

31. Binnig G, Rohrer H, Gerber C, Weibel E (1983) 7 × 7 reconstruction on Si(111) resolved in real space. Physical Review Letters 50:120.

32. The reconstruction is on the silicon (111) surface and is referred to as a 7 × 7 reconstruction because the surface unit cell has 7 atoms on each side.

4

Made to Measure

By the end of the Twenties, there was a new science. One based on uncertainty and probability. The pioneers of quantum mechanics including de Broglie, Heisenberg, Schrödinger, and Pauli, were responsible for a scientific revolution that changed our concept of the atom giving us a new way of looking at, and thinking about, the world around us. By the end of the Thirties, we had commercially available electron microscopes that enabled imaging of the structure of matter with a previously unparalleled resolution. Electron microscopes worked so well because wave-like electrons can be accelerated to speeds so fast that their wavelength becomes a tiny fraction of that of visible light. So small, in fact, that it is less than the distance across a single atom.

Through an electron microscope scientists saw for the very first time the arrangement of atoms in a solid. Images of line defects that were named dislocations by British physicist and mathematician Geoffrey Taylor in 1934 proved that a late nineteenth century invention to account for the permanent deformation of metals was correct. Watching dislocations move in the microscope enabled an explanation of how metals can be strengthened by the process of work hardening. Ancient metalworkers knew that repeated hammering of metals would make them harder. Electron microscopes provided the supporting scientific evidence of how work hardening actually worked. And electron microscope images showed the processes by which crystals nucleate and grow.

© The Author(s), under exclusive license to Springer Nature
Switzerland AG 2023
M. G. Norton, *A Modern History of Materials*,
https://doi.org/10.1007/978-3-031-23990-8_4

By the end of the Forties, there was a new microelectronic technology—the transistor—based on the semiconductor element germanium. At the beginning of the Fifties a process accidently discovered to produce thin single crystal wires of tin would be applied to grow much larger crystals of germanium and then silicon, which in a short amount of time ushered in the integrated circuit, the ubiquitous "silicon chip". Eventually there was an interconnected world through an Internet of Things and a social network all using billions of transistors formed on individual wafers of silicon.

In the past century, three methods of making materials have been developed that have accelerated the technological changes that have come to define our world: pulling single crystals of silicon from the melt, drawing ultra-pure glass optical fibers, and blowing polyethylene sheet. A fourth method, 3D printing, may have similar potential for change.

There is abundant evidence of crystal growth in nature. For instance, single crystals of quartz weighing over 40 tons have been found. These formed as magma cooled and transformed into a solid.[1] A monstrous crystal of the mineral beryl from Malakialina in Madagascar weighing in at over 400 tons is perhaps the largest example of nature's ability to form crystals. Even just one hundred years ago the field of synthetic crystal growth was very much in its infancy. As it turned out, synthesizing single crystals—particularly large ones—proved to be a very challenging engineering problem. Early pioneers including French chemist Auguste Verneuil, considered the father of crystal growth, were only able to produce very small globular single crystals from which they got the name "boules". A typical Verneuil-grown crystal of sapphire or ruby is about 125 carats or 25 g. Midway through the twentieth century, however, it was clear that there was a need to be able to grow large, high-purity, defect-free crystals.

The first transistor—a point-contact transistor—constructed and tested in 1947 by Bell Labs physicists John Bardeen and Walter Brattain used a small crystal of germanium as the active semiconducting material.[2] Germanium was also the material used by another Bell Labs physicist, William Shockley, to construct a more reliable modification of the point-contact transistor called a junction transistor. While the potential of the transistor was not appreciated by the popular news media, it was quickly recognized by those working in and reporting on the field: " … [it] is destined to have far-reaching effect on the technology of electronics … " exclaimed the September 1948 issue of *Electronics* magazine. Almost a decade after their groundbreaking research recorded meticulously in the pages of their Bell Labs notebooks Shockley, Bardeen, and Brattain shared the 1956 Nobel Prize in Physics for "their discovery of the transistor effect."[3] But it was not the germanium transistor

that was to have the predicted "far-reaching effect". It was silicon that was to enable an information and communication revolution. By 1966 the number of silicon transistors sold exceeded 480 million units, by-passing, for the first time, sales for those made using germanium.[4]

An early challenge for both the point-contact transistor and the junction transistor was the lack of pure, uniform semiconductor materials. Bell Labs chemist Gordon Teal pointed out that to make more reliable devices large single crystals of germanium and silicon would be required. Despite finding little support from the lab's management Teal set about with the help of mechanical engineer John Little and technician Ernest Buehler to grow the much needed crystals. The technique that Teal, Little, and Buehler decided upon was one that had been first demonstrated—albeit accidently—over thirty years previously by Polish chemist Jan Czochralski.

Czochralski was born in 1885 in Kcynia, a small town in what was then part of the Prussian Province of Posen. While still a young man he moved to Berlin and in 1907 joined the company Allgemeine Elektrizitäts Gessellschaft (AEG), which at the time was the largest metal processing factory in Germany. Czochralski rose through the ranks to become chief of AEG's metals laboratory in 1914. One of his interests in this new leadership role was studying the crystallization of metals: their transformation from a hot liquid melt into a crystalline solid.

Being able to control the process of solidification was recognized as important almost as far back as our earliest uses of metals where liquid arsenic-bronze alloys were cast into molds to form hardwearing swords and daggers. Arsenic was eventually and deliberately replaced by tin, which led to the dawn of the Bronze Age. The speed of solidification of a bronze melt can be judiciously used to change not only the structure of the resulting metal but its properties. Rapid solidification generally produces smaller grains, which can lead to a higher strength alloy. As early as 1200 BCE rapid cooling was used to strengthen iron by the formation of the very hard martensite phase. Chinese metalworkers who were experts with casting iron used the technique to form enormous structures such as the impressive hollow-cast 40-ton Cangzhou lion in Hebei province. Dating from 930, the magnificent lion is the largest iron casting artwork in the world.

Despite the widespread use of solidification over thousands of years and the development of an empirical understanding that went along with it, the accompanying science has a much more recent history. Modern solidification science can be traced back to the 1940s, when engineers and scientists began to make use of mathematical models to describe what was happening during cooling of a melt. But it was not until the mid 1950s and the pioneering work

of British-born Harvard metallurgy professor Bruce Chalmers, that solidification became an engineering science and the necessary conditions under which a liquid transforms into a solid were determined.[5] In particular, Chalmers was interested in the mechanism by which atoms and molecules moved across the boundary from the disordered liquid phase into the regular arrangement of a crystalline solid as the temperature is lowered.

So, in 1914 the field of solidification was largely empirical, lacking the detailed scientific interpretation that would come several decades later. Jan Czochralski's work in determining how rapidly metals crystallized was unique and would gain particular significance in the history of crystal growth. It was during these crystallization experiments using the low melting point metals tin, zinc, and lead that Czochralski made his most fortuitous accidental discovery.

The story of that discovery is recounted as follows: "A crucible containing molten tin was left on his table for slow cooling and crystallization. Czochralski was preparing his notes on the experiments carried out during the day when at some point, lost in thought, he dipped his pen into this crucible instead of [the] inkwell placed nearby the crucible. He withdrew it quickly and saw a thin thread of solidified metal hanging at the tip of the nib. The discovery was made! He had generated a phenomenon never occurring in Nature—crystallization by pulling from the surface of a melt."[6] The crystallized wire just a few millimeters wide, but many centimeters long, was a single crystal.

The paper describing the crystal growth technique, which is frequently referred to simply as the CZ method, was subsequently published in 1917.[7] Within a year of its publication Czochralski's crystal pulling method was used by Hans von Wartenberg at the Technical University of Gdansk to grow single crystal wires of zinc onto oriented "seed" crystals.[8] The seed acts as a template ensuring that the growing crystal maintains the desired orientation, working in much the same way that Lego® blocks neatly stack one upon the other in a predetermined way. Using a small crystal of the material to be grown to "seed" crystallization by pulling from the melt is the approach that is still used to this day over hundred years since Czochralski's discovery of crystal pulling and von Wartenberg's use of seed crystals.

The Bell Labs partnership of Gordon Teal and John Little pulled their first single crystal of germanium on October 1, 1948, by slowly withdrawing a seed crystal from a melt of very pure germanium at a temperature of 938°C.[9] Soon thereafter Teal and Ernest Buehler reported at the meeting of the American Physical Society held in Washington D.C that they had successfully grown single crystals of silicon also by the process of pulling from

the melt. On June 29, 1951, Buehler and Teal filed a United States patent for a process describing a method of producing "semiconductive crystals of uniform resistivity" in June 1951.[10,11]

Following the work of Teal, Little, and Buehler, the Czochralski method became the most important technique for the production of large single crystals. It is used to grow crystals of a wide variety of materials including synthetic gemstones of ruby, sapphire and garnet, crystals of compound semiconductors such as gallium arsenide and indium phosphide, but none more important than its signature application, the growth of silicon single crystals.

To grow silicon crystals using the crystal pulling method a high-purity silicon seed crystal of the desired orientation is lowered into a silica crucible containing the silicon melt at a temperature above about 1420°C. (Silicon melts at 1414°C.) By reducing the temperature slightly, silicon atoms in the liquid begin to attach onto the seed in such a way that the crystal orientation is maintained. As the seed is raised the growing crystal forms a narrow "neck" that will end up supporting the entire weight of the crystal. Growth continues as more and more of the atoms in the melt join the growing crystal as it is slowly raised from the melt. For single crystal silicon a typical growth rate—how fast the crystal is pulled from the melt—is between 100 and 200 mm per hour. This is very much slower than the growth rates used by Czochralski for his metal crystals, which were grown at speeds of about 100 mm per minute.[12] Because silicon is highly reactive with oxygen—which accounts for the abundance of silicate minerals in the Earth's crust—the entire single crystal growth process is carried out in a shroud of argon gas.

In 1952, Buehler and Teal reported that they had grown "single crystals of silicon having a high degree of lattice perfection and chemical purity". The crystals were 5 inches in length and had a diameter up to 1 inch (25 mm).[13] By 1956, 4-inch diameter boules were produced. Three years later crystals with diameters as large as 6 inches were possible.

In addition to requiring single crystals of high purity the microelectronics industry requires that the crystals are free from dislocations—line defects that disrupt the regular arrangement of atoms. For many materials even large numbers of dislocations do not present any problems. In fact in metals, as noted earlier, they are often desirable. However, in a semiconductor they must be avoided entirely because breakdown in silicon p-n junctions—the basic building blocks of all transistors—was found by Bell Labs scientists A.G. Chynoweth and G.L. Pearson to occur preferentially at dislocations.[14] Charles Frank at the University of Bristol in England and William Thornton Read, Jr. of Bell Labs had already shown that dislocations in silicon could

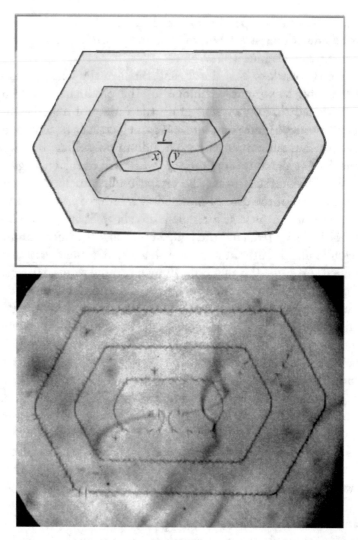

Fig. 4.1 Illustration of how dislocations can multiply (upper). Under an applied stress new dislocations form as shown in the infrared image (IR) image (lower) of a Frank-Read dislocation source in silicon (Reprinted from Carter CB, Norton MG (2013) Ceramic Materials: Science and Engineering 2nd ed. Springer, New York, p 224. Original image by W.C. Dash)

multiply when the crystals were put under an applied stress as shown in the image in Fig. 4.1.[15]

To interpret what is happening, suppose a dislocation is pinned at two points, x and y, that are a distance l apart. Under an applied stress the dislocation, which appears as a dark line in the microscope image will bow out. Eventually the two segments x and y will meet annihilating each other to form a complete loop and reforming the pinned dislocation. The completed loop expands away from the original source and the process continuously repeats itself generating more and more dislocations until such time that the applied force is removed. Recognizing the negative impact of dislocations on the performance of semiconductors and realizing that they could multiply made it essential to find a way to ensure they were not present in the growing boule to begin with.

William Dash a crystal grower at General Electric Research Laboratories in Schenectady, New York solved the dislocation problem in 1959. Dash modified the Czochralski process to enable growth of silicon crystals that were completely free of dislocations.[16] His approach was remarkably simple, but it required a detailed understanding of how dislocations in silicon move. This movement as was mentioned in Chapter 3 is referred to as slip. For each crystal structure there are preferred slip planes along which dislocations move. The preferred slip planes are usually the ones that contain the largest number of atoms and are the closest together.

During the initial stages of Czochralski growth as the seed was being withdrawn Dash reduced the diameter of the neck region by pulling the crystal really quickly, allowing the dislocations to slip the short distance to the surface of the hot crystal where they literally vanish. Once all the dislocations had been removed the growth rate was slowed, the neck widened until a boule of the necessary diameter, free of dislocations, was produced.

After cooling the finished boule is sliced using high-speed diamond-tipped saws into individual wafers about the thickness of a dime. Cutting leaves behind a rough surface so the wafers must be ground and polished until they are mirror smooth. It is into these polished wafers that transistors and other circuit components that make a completed integrated circuit, or silicon chip, are formed.

With the development of a process to grow large high-purity dislocation-free single crystals of silicon it became possible to mass produce silicon wafers, now as big as 18 inches (~450 mm) in diameter, as a substrate for the fabrication of billions of transistors. About 95% of all single crystal silicon is now produced by the Czochralski method, a method that came from a fortunate accident in a metal's laboratory in Germany.

Being able to pull wider and wider boules from the melt allowed more and more transistors to be formed simultaneously into a single wafer providing an economy of scale that has made the semiconductor industry famous. For instance, the computer memory (DRAM) industry is a huge, multi-billion dollar per year business.[17] Yet as memory power has *increased*, costs to the consumer have gone *down* because more chips can be fabricated in larger

Fig. 4.2 Top "seed" end of a Czochralski grown silicon crystal. At this point the crystal is about 3 mm on diameter

wafers than in smaller wafers. As an example of this efficiency of scale, over two hundred 16 GB DRAM chips can be formed in a single 400-mm (16 inch) wafer. That is four times as many as can be fabricated in the same number of processing steps in a 100-mm diameter wafer. For the ubiquitous smartphone such as the iPhone and the Samsung Galaxy the average DRAM content is 4 GB. Over two hundred and fifty 4 GB DRAM chips can be formed in a 400-mm wafer compared to just over one hundred in a 300-mm (12 inch) diameter wafer.[18]

Figure 4.2 is a picture of the seed end of a Czochralski silicon boule. The complete crystal has a diameter of 300 mm and weighs more than 250 kg. The narrow neck region, which is only 3 mm in diameter and 30 mm long supports the entire weight of the growing crystal. The striations—growth rings—which form along the length of the crystal are a common feature in Czochralski grown silicon. They usually result from slight temperature fluctuations occurring at the interface between the liquid and solid.

Before Gordon Teal and his colleagues at Bell Labs had adapted Jan Czochralski's crystal pulling method to grow single crystals of silicon and germanium there were other ways to produce the germanium crystals needed for the early transistors. The most important of these processes was developed and patented by Harvard professor Percy Williams Bridgman.[19]

Bridgman's patent was filed on February 16, 1926 and awarded on February 24, 1931. It describes a process very different from crystal pulling.

In Bridgman's method a crucible containing a melt of the crystal to be grown is lowered at a speed of about 20 mm per hour through a long vertical tube furnace, which is held at two different temperatures. The top half of the furnace is hot enough to maintain the melt, while the bottom half is kept below the melting temperature of the crystal. As the crucible enters the cooler region of the furnace the melt begins to solidify. If the tip of the crucible has the appropriate curvature and the cooling rate is slow enough large single crystals can be grown using the Bridgman method. However, even the largest Bridgman crystals are much smaller than those grown by the Czochralski technique. And it is not possible to guarantee that the Bridgman-grown crystals are dislocation free.

Among the first crystals to be grown using Bridgman's method were the low melting temperature metals lead, tin, zinc, and cadmium. These are some of the same materials that were studied by Jan Czochralski in his crystal growth experiments. The Bridgman method has also been used to grow single crystals of the important compound semiconductor gallium arsenide, which is used in high-speed computer chips and is a component in highly efficient solar cells. Recently single crystals of cadmium telluride exceeding 1 kg in size have been grown using a high-pressure modification of the Bridgman method.[20] Cadmium telluride is used for X-ray and gamma ray detectors that are important for both defense applications and in medical imaging. But the many advantages of the Czochralsi process led to the eventual decline in the use of Bridgman's method.

Czochralski-grown silicon is among the purest and structurally most perfect materials ever produced. Impurity atom concentrations are as low as 1 impurity per 100 billion silicon atoms (a purity of 99.999999999 percent). And using Dash's method the boules are free from any crystalline defects larger than the size of an individual atom.[21]

Large high-purity and high-quality single crystals of silicon have without doubt been the enabling material for the multitude of electronic devices used in our everyday lives including laptops, tablets, smartphones, and all the associated technology that combines to form the Internet of Things. Although research is currently ongoing and several different technologies are being examined, it is possible that silicon may also turn out to be the best material at the heart of next generation computers—quantum computers.[22]

Researchers around the world are exploring several different silicon architectures, using Czochralski grown single crystals, that might be used as the building blocks for quantum computers. A more detailed examination of the materials behind quantum computers is the topic of Chapter 8, but here is one example proposed by Kohei Itoh a professor at the Keio University

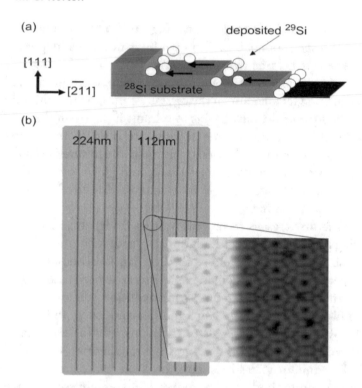

Fig. 4.3 (a) An illustration of how single atom wide silicon-29 (^{29}Si) wires can form at step edges on a silicon-28 (^{28}Si) substrate. (b) A scanning tunneling microscope image of the surface of the silicon substrate. The right figure is a close-up of the step-edge structure (Reprinted from Itoh KM (2005) An all-silicon linear chain NMR quantum computer. Solid State Communications 133:747-752 with permission from Elsevier)

Quantum Computing Center in Yokohama, Japan. Figure 4.3 shows Itoh's proposed structure.

A single crystal silicon substrate is cut from a larger Czochralski grown wafer. The direction of the cut does not coincide with a low energy plane, for instance the slip plane, in the silicon structure. So, when heat treated at a temperature of 1260 °C for a few seconds in vacuum the surface atoms realign themselves forming a lower energy configuration consisting of a series of atomically straight steps. We saw earlier in Chapter 3 in the description of the scanning tunneling microscope another example of how silicon atoms can rearrange on the material's surface. Atoms of a silicon isotope (silicon-29) are deposited onto the substrate under ultra-high vacuum conditions where they align preferentially along the step edges forming atomically straight nanowires. Each silicon-29 atom is an individual *qubit*.[23] (More about qubits in Chapter 8.) The lower image in Fig. 4.3 is a scanning tunneling microscope image showing the arrangement of atoms on the surface and the atomically straight step edges formed on the single crystal silicon surface.

A technology as important for our high-speed digital communication as the silicon chip itself is the optical fiber. The performance of an optical fiber, like that of the silicon chip, relies on its purity. Any defect that absorbs the light as it passes along the length of the fiber will cause a loss in the signal intensity reducing the amount of the incident light that arrives at the other end of the fiber. To be a success optical fiber technology requires defect-free glass with an unparalleled clarity.

The idea of sending optical signals along hair-like glass fibers was first proposed by Charles Kuen Kao and George Hockham in a paper published in 1966.[24] At the time both Kao and Hockham were researchers at Standard Telecommunication Laboratories (STL) in the United Kingdom. The impact of this original paper combined with Kao's later work was so great that he was awarded the 2009 Nobel Prize in Physics "for groundbreaking achievements concerning the transmission of light in fibers for optical communication."[25] The Nobel Foundation went on to underscore the significance of Kao's work: "The rapid transmission of signals over long distances is fundamental to the flow of information in our time."

One of the early experiments using glass to guide light took place in 1930 when German American physician Heinrich Lamm created a flexible endoscope designed to look around the internal twists and turns of the body, which the rigid variants of the time would not allow. Although Lamm's bundle of thin glass fibers transmitted an image, it was not of any previously inaccessible part of the human body, but rather of a light bulb filament. Perhaps not surprisingly the image was of a very poor quality as the glass fibers available to Lamm and others working in this area at the time were riddled with defects and only able to transmit light over very short distances. The poor quality of early glass lenses also limited the resolving power of microscopes. Long-distance information transfer would require significantly better quality fibers. Ones where imperfections in the glass would be minimized and light would not be lost shortly after it left the source.

Light is transmitted from one end of an optical fiber to the other through the process of total internal reflection (TIR). Assuming that the light is not absorbed or scattered by defects in the glass it will be repeatedly reflected from one internal surface to another along the entire length of the fiber. Kao and Hockham determined that for glass fibers to be used to transmit optical communications the attenuation—or loss—of the signal must be less than 20 decibels per kilometer. In other words, at least 1% of the incident light remains in the fiber after it has traveled 1 km (0.6 mile). At the time of Kao and Hockham's proposal, fibers with losses of 1,000 decibels per kilometer or even higher were commonplace. These losses were a long way from meeting the performance required for optical fiber communication.

In order to make a medium that would be perfect enough to transmit light over long distances Kao and Hockham recognized that a silica glass could be used providing any impurities, particularly transition metal elements such as iron, copper, and manganese were reduced down to parts per million or even parts per billion. This level of purity is equivalent to having just one impurity atom for every billion host atoms: a purity approaching that required for semiconductor silicon. To achieve these unprecedented purities conventional glass processing using readily available raw materials would not be suitable. A brand new technology was needed, one that would produce a glass where the composition could be very tightly controlled.

In 1973, Bell Labs, an organization that played such a major role in the history of the transistor and the growth of silicon crystals, developed a process for making ultra-transparent glass that could be mass produced into a low-loss optical fiber.[26] Almost fifty years later the technique known as modified chemical vapor deposition (MCVD) remains the standard for manufacturing optical fiber cables today.

Rather than forming glass in the traditional way, mixing oxide or carbonate powders together, melting them, and then cooling the hot liquid, the glass for an optical fiber is made by reacting ultra-high purity gases together and forming the glass, layer-by-layer, inside a hollow silica tube. Combining gases as the precursors to form the glass rather than mixing impure powders was the key innovation that allowed the formation of a highly transparent material.

Modified chemical vapor deposition starts with a high purity silica glass tube—a preform—6 mm thick with an inside diameter of 19 mm. Reactant gases, typically mixtures including silicon tetrachloride ($SiCl_4$) and germanium tetrachloride ($GeCl_4$), are fed in at one end of the hot preform, which is heated by a flame similar to that used in oxy-acetylene welding. The flame can travel back and forth along the length of the preform. At temperatures of more than 1300 °C, the gases react with high-purity oxygen to deposit a glassy oxide layer—a mixture of silicon dioxide and germanium dioxide—on the inside wall of the preform. About 1 g of glass is formed per minute. The composition of the glass—the ratio of silicon to germanium atoms—can be modified through the thickness of the layer by depositing more or less of one component. As the composition of the glass is varied there is a gradual change in the refractive index (how much the light is bent as it travels along the fiber) allowing the path of the light to be carefully controlled to decrease losses and speed up transmission through the fiber. Up to 100 individual layers may be deposited within the preform. Each layer having a slightly different composition.[27]

Once a sufficient glass thickness has been built up, the tube is collapsed by heating to an extremely hot 2000°C. Glass fibers are then drawn from the collapsed preform at a rate of more than 10 m per second. The resulting fiber consists of a high-purity silica glass cladding made from the original preform with a core having a layered glass composition designed to minimize the time it takes light to travel along the fiber.

Just over twenty years after Kao and Hockham's landmark publication, optical fibers were crossing the Atlantic Ocean and another ten years later spanning the mighty Pacific. By the end of the twentieth century much of the world's telecommunications was travelling through clear glass fiber optic cables. In the year Kao's Nobel Prize was awarded in Stockholm more than one billion kilometers (about 600 million miles) of optical fiber had been deployed around the world. Enough to circle the globe more than 25,000 times.

In an interview with Radio Television Hong Kong, Charles Kao stated: "I cannot think of anything that can replace fiber optics. In the next 1000 years, I can't think of a better system."[28]

We and everything around us are comprised of atoms, the building blocks of matter. These atoms combine to form molecules including proteins, enzymes, the nucleic acids DNA and RNA, and polysaccharides such as cellulose and starch. All these molecules that are essential for life. In the early twentieth century there was a debate within the scientific community about how large these molecules could become. Hermann Staudinger an organic chemist working first at the Federal Institute of Technology in Zürich and then at Freiburg University in Germany claimed that they could be very large indeed, amounting to tens or even hundreds of thousands of atoms in size. Staudinger's work, which began in 1920 showed how small molecules with just a handful of atoms can join to form long chains to become very large molecules, which we call macromolecules. Combining small molecules together in ways that are not found in nature was the basis for the development of a whole class of synthetic materials known as polymers or plastics. For instance, polystyrene consists of long chains of styrene molecules. Each chain comprising over 100,000 atoms.[29]

During his prolific career Staudinger wrote approximately 500 papers. Many on these were on the natural polymer cellulose. Fifty papers were on rubber, where there was growing interest in producing synthetic versions that would eliminate the reliance on the natural material. The impact of Staudinger's work led to his award of the 1953 Nobel Prize in Chemistry "for his discoveries in the field of macromolecular chemistry."[30] These discoveries recognized by the Swedish Academy of Sciences marked the beginning of an era of molecular design where a whole host of polymers were synthesized: polystyrene (1929), polyester (1930), polyvinylchloride (1933), polyethylene (1933), nylon (1935), Teflon (1938), and polyethylene terephthalate (1941). Each of these materials possesses unique properties that have enabled their use in a wide and ever-increasing range of applications. But one feature they all share is that none of these polymers has any natural analogs. They are purely creations of human ingenuity.

Advances in chemistry, particularly in the discovery of new catalysts, were needed to realize many of the polymers envisaged by Staudinger and others and to turn these materials into commercial products. Commercialization required not only an economical synthesis method but also the development of innovative ways to shape and form polymers into sheets, tubes, wires and a myriad of other shapes.

Structurally and chemically the simplest of all polymers is polyethylene: a macromolecule consisting of long chains of ethylene molecules (C_2H_4). Polyethylene has become one of the most widely used plastics in the world and, perhaps not surprisingly, it is the most abundant to be found in municipal solid waste dumps. Dow Chemical Research Fellow Mehmet Demirors describes how widespread the uses of polyethylene are and how it is so interwoven with many aspects of our daily lives impacting everything from the safe delivery of electric power into our homes to protecting us from food-borne bacteria and other harmful pathogens.[31] As a coating, polyethylene insulates miles and miles of power lines and electricity cables. As a thin film, polyethylene wrap may extend the shelf life of perishable food items, reducing waste. Polyethylene foams help protect us from accidents and injuries in our cars and on the football field. And while polyethylene may be one of the biggest culprits in filling our landfills, thick polyethylene sheets in sanitary landfills can prevent leakage of dangerous chemicals into the groundwater.

When drawn into pipes polyethylene is responsible for efficiently transporting almost all of the natural gas in the United States. But by far the largest use of polyethylene is in the form of transparent thin films, which was one of the suggested uses for polyethylene in a list made by Imperial Chemical Industries (ICI) researchers way back in 1936. However, the suggested application was not for food packaging but for covering airplane wings because the material has "high water resistance and good low temperature properties."[32]

To make the large sheets of polyethylene film that are used in packaging, an extrusion process was developed that in some ways is akin to glass blowing, except it occurs at a much lower temperature and on a much larger scale.

Film manufacturing starts with melting down small polyethylene pellets, nurdles, at a temperature around 110°C. The molten mass is pushed through a circular die to form a continuous tube called the "bubble." As air is blown into the bubble it expands until it reaches the desired diameter. The bubble is then drawn vertically up a tower, which can be several stories high. As the bubble rises it cools before being flattened to form a film. Film thickness is controlled by the speed at which the bubble is pulled from the die. Depending on the application the film can be anywhere from less than 10 μm up to as much as 75 μm thick. Even at its thickest the film is still considerably thinner than a human hair. Saran is a trade name for a widely used polyethylene food wrap that is about 12 μm thick. The width of the film is controlled by the amount of air blown into the bubble. Blown films represent by far the largest market segment of the more than $16 billion, and growing, plastic film market in the United States.

In the Apple TV + science fiction series *For All Mankind* NASA establishes the *Jamestown* base on the Moon on October 12, 1973, during the presidency of Richard Nixon. Later the following year the USSR announce the establishment of their own Moon base that they named *Zvezda*, "star" in Russian. Both bases are able to accommodate large numbers of personnel living on the Moon for many months on end with supplies routinely sent up from Earth. These bases are a work of fiction existing in the minds of the show's writers and on our viewing screens. In reality, a total of just twelve human beings have set foot on the Moon. The last lunar landing mission was almost fifty years ago when the astronauts of Apollo 17 spent a total of 75 h on the Moon's surface in December 1972. During this mission over 100 kg of lunar material were collected. With the Artemis program, which was launched in December 2017 NASA plans to land the first woman and first person of color on the Moon by 2024. Artemis is part of an even bolder goal to establish a sustainable base at the South Pole of the Moon by the end of the decade and eventually begin planning for the 140 million-mile journey to our neighbor Mars.[33]

Establishing a sustainable crewed base on the Moon requires many things. At the top of that list are a source of power and access to water. Because they are bathed in nearly continuous sunlight—peaks of eternal light—the polar mountains are suitable locations for setting up arrays of solar cells to provide electrical power to the base. Exploration of the lunar surface over the past two decades has detected ice in the permanently shadowed craters around the Moon's poles. Recent exciting research from NASA's Stratospheric Observatory for Infrared Astronomy (SOFIA) confirms that the presence of water may be more widely distributed than previously thought and not limited solely to the Moon's cold shadowed recesses.[34] In order to allow the water to remain exposed on the lunar surface the researchers suggest that one possibility is that it is stored within glass beads formed by the heat generated by the impact of micrometeorites.

In addition to considering the dual requirements of power and water, a number of research groups have been looking at the possibility of using lunar regolith—rock fragments and powdery Moon dust—as a raw material in the same way we use minerals in the Earth's crust. Lunar regolith might be turned into a variety of useful components fabricated directly on the Moon rather than relying on shipments coming from Earth. There are very ambitious proposals for using lunar regolith including major infrastructure projects such as shields to protect astronauts against the twin dangers of meteoroids and cosmic radiation and towers for mounting solar panels to capture the maximum amount of sunlight. Eventually, the goal is to establish a manufacturing capability on the Moon sourced with locally available raw materials.

Because of the limited quantities of samples collected from the Moon researchers have to make use of regolith analogs that simulate, as closely as possible, the composition of actual lunar material. Table 4.1 compares the composition of a commercial lunar regolith simulant, an actual regolith sample collected from the Moon's surface, and the Earth's crust. By far the largest component on Earth and on the Moon is silicon dioxide (silica), which is present in a wide range of silicate rocks such as feldspars, for instance anorthite ($CaAl_2Si_2O_8$), and pyroxenes including the mineral enstatite ($Mg_2Si_2O_6$).

A research team led by engineers at Loughborough University in the United Kingdom demonstrated that a 3D printing technique known as selective laser melting (SLM) could be used to consolidate simulated Moon rock into solid test pieces.[37] Three dimensional structures were built up layer upon layer by selectively melting and resolidifying regolith powder using a high-power laser. The powder contained particles between 20 to 50 μm in diameter. Each individual layer was one to two particles thick. The shape of the final object is determined by a detailed design stored on a computer, which controls the deposition of the powder layers and the location of the laser beam as it sweeps across the growing component.

In the same study, the researchers found that simulated Martian regolith—there are no actual mineral samples from the surface of Mars—was not suitable for 3D printing using the selective laser melting technique. One of the differences between simulated lunar regolith and Martian regolith is the Martian variety has a much higher melting temperature; 1330°C compared

Table 4.1 Composition of Lunar regolith simulant, actual Moon rock, and the Earth's crust

Chemical compound	Lunar regolith simulant[35]	Lunar regolith—actual sample (14163)[35]	Earth's crust[36]
Silicon dioxide (SiO_2)	46–49	47.3	60
Aluminum oxide (Al_2O_3)	14.5–15.5	15	15
Calcium oxide (CaO)	10–11	10.4	5
Magnesium oxide (MgO)	8.5–9.5	9	5
Iron oxide (FeO)	7–7.5	9	5
Ferric oxide (Fe_2O_3)	3–4	3.4	4
Sodium oxide (Na_2O)	2.5–3	2.7	3
Titanium dioxide (TiO_2)	1–2	1.6	0.5
Potassium oxide (K_2O)	0.75–0.85	0.8	3
Diphosphorus pentoxide (P_2O_5)	0.6–0.7	07	0.2
Manganese oxide (MnO)	0.15–0.20	0.2	0.1
Chromium oxide (Cr_2O_3)	0.02–0.06	–	–

to 1150°C. Because of their high melting temperatures 3D printing of these materials is more challenging than for low melting temperature metals and, in particular, plastics where the technique has become routine.

Figure 4.4 shows an example of a foam structure that could be imagined to form part of the external walls of a lunar outpost. This section, which weighs over 14 kg was 3D printed from regolith simulant using a technique called fused deposition modelling or FDM. In fused deposition modelling of structures made from powders such as the simulated lunar regolith an "ink" comprised of the finely-ground powder mixed with a cocktail of plasticizers and organic liquid is extruded through a narrow nozzle. Deposition speeds vary depending on the mixture and the complexity of the structure being built but are typically in the rage of 5 to 120 mm per second. Fabrication of large 3D printed objects on the Moon is full of obstacles not least of which is that to print large objects you need a large quantity of supplies, which will make the process incredibly, maybe prohibitively, expensive. A quick calculation illustrates the challenge. A typical modest single-family single-story home weighs about 180,000 kg or 400,000 pounds. It costs $4,000 to $13,000 per kilogram to transport supplies from a spacecraft launched from the ground to Low Earth Orbit. So, transporting the necessary raw materials from the Earth to the Moon would cost around $2 billion! That is for just one building.

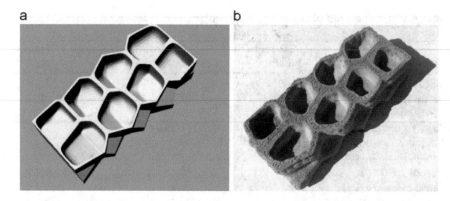

Fig. 4.4 Possible building component for an outpost on the Lunar soil by means of a novel 3D printing technology. On the left is the computer-generated drawing of the component. On the right is the actual printed block, which weighs 14.4 kg (Reprinted from Cesaretti G, Dini E, De Kestelier X, Colla V, Pambaguian L (2014) Building components for an outpost on the Lunar soil by means of a novel 3D printing technology. Acta Astronautica 93:430-450 with permission from Elsevier)

While constructing buildings on the Moon using 3D printing is still a very long way off, even for the most optimistic projections, the technique is being used to build homes here on Earth. As just one example, Palari Homes and Mighty Buildings are collaborating to build a number of new houses across the state of California using 3D printed panels.

To demonstrate the flexibility of 3D printing, a group at the Politecnio di Milano in Italy have used a 3D printing method called liquid deposition modeling (LDM) to fabricate nanocomposite microstructures containing carbon nanotubes.[38] A picture of their conductive nanocomposite woven structure, forming part of a simple electrical circuit, is shown in Fig. 4.5. The composite consists of multi-walled carbon nanotubes and poly(lactic acid), a biodegradable polymer, delivered in a liquid slurry made using dichloromethane.

SLM, FDM, and LDM are all forms of 3D printing where a 3D object is fabricated layer-by-layer from a digital model. Collectively these methods come under the umbrella of additive manufacturing, finding use in industrial sectors ranging from aerospace, to automotive, to bioengineering. While the traditional applications for 3D printing were primarily for rapid prototyping, the benefits of additive manufacturing—design flexibility, ultimate customization, waste minimization, and the ability to manufacture complex shapes—have made it one of the most widely studied techniques for materials fabrication.

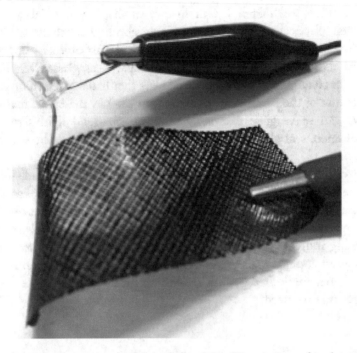

Fig. 4.5 A conductive polymer fabric formed by 3D printing of polymer/carbon nanocomposites via the technique of liquid deposition modeling (Reprinted from Postiglione G, Natale G, Griffini G, Levi M, Turri S (2015) Conductive 3D microstructures by direct 3D printing of polymer/carbon nanocomposites via liquid deposition modeling. Composites Part A 76:110-114 with permission from Elsevier)

Our ability to shape materials has enabled a continually evolving range of applications that have shaped our world and, in some cases, defined us. Our early ancestors shaped flint to make tools that created an evolutionary advantage. Over time the tools became increasingly complex and refined allowing them to be tailored for specific tasks. For instance, a spear would be shaped very differently from a stone awl.

The Egyptians were foremost among ancient peoples in their mastery of manufacturing techniques. They mastered the earliest example of a layer-by-layer technique called core forming where a molten glass is wound around a solid core supported by a rod. After the vessel cooled, the core was removed leaving behind a hollow container. The most delicate of which would be used to hold prized scented oils. Conceptually, core forming shares many similarities to the now ubiquitous and inexpensive 3D printing techniques without having the advantage of computer control.

The Romans were familiar with glass blowing, a technique that works in a similar way to how we make enormous thin sheets of plastic by blowing air into the hot soft material. Our history of shaping materials is not only about their physical form, but increasingly about their chemical composition. While the Romans could produce beautiful glass objects—bowls, vases, jugs—by glass blowing they struggled to make clear glass. Even the most skilled of ancient glassworkers produced glass that had a greenish or yellowish tinge because of the presence of impurities in the raw materials, mainly iron. Now we can make glass that is so pure we can send light signals along many kilometers of glass optical fibers without it being lost by scattering or absorption.

Controlling composition has allowed us to grow single crystals of silicon that were critical in enabling a technological revolution that has changed everything from how we make things to how we communicate. In the next chapter we will see an example of how we can make materials by moving and placing individual atoms one at a time. We will also see how many of the proposed applications for nanomaterials, for instance carbon nanotubes and graphene, are limited by our inability to produce them in the forms and quantities that we need.

Notes

1. Rockwood PC (1981) The largest crystals. American Mineralogist 66:885–907.
2. Bardeen J, Brattain WH (1948) The transistor, a semi-conductor triode. Physical Review 74:230–23. Walter Brattain's logbook describing the experiment was dated December 1947. The journal paper appeared the following year. The germanium was provided by JH Scaff and HC Theuerer.
3. The Nobel Prize in Physics 1956. NobelPrize.org. Nobel Media AB 2021. Fri. 12 Feb 2021. https://www.nobelprize.org/prizes/physics/1956/summary/.
4. In 1966 481 million silicon transistors were sold compared to the 369 million germanium variety. Data from Electronic Industries Association Yearbook (1967).
5. Jackson KA, Chalmers B (1956) Kinetics of solidification. Canadian Journal of Physics 34:473–490.
6. Tomaszewski PE (2002) "Jan Czochralski—father of the Czochralski Method. Journal of Crystal Growth 236:1–4.
7. Czochralski J (1917) A new method for the measurement of the velocity of crystallization of metals. Zeitschrift für Physikalische Chemie 92:219–221.
8. von Wartenberg H (1918) Verh. Dtsch. Phys. Ges. 20:113.
9. Teal GK, Little JB (1950) Growth of germanium single crystals. Physical Review 78:647.

10. Teal GK, Buehler E (1952) Growth of silicon single crystals and of single crystal silicon p-n junctions. Physical Review 87:190.
11. Buehler E, Teal GK (1956) Process for producing semiconductive crystals of uniform resistivity. United States Patent 2,768,914.
12. Uecker R (2014) The historical development of the Czochralski Method. Journal of Crystal Growth 401:7–24.
13. Buehler E (2004) 50 years progress in crystal growth in: Feigelson RS (ed.) Elsevier, Amsterdam, Tokyo p. 105.
14. Chynoweth AG, Pearson GL (1958) Effect of dislocations on breakdown in silicon p-n junctions. Journal of Applied Physics 29:1103.
15. Frank FC, Read Jr WT (1950) Multiplication processes for slow moving dislocations. Physical Review 79:722.
16. Dash WC (1959) Growth of silicon crystals free from dislocations. Journal of Applied Physics 30:459–474.
17. DRAM is the acronym for Dynamic Random Access Memory. These integrated devices store data in a computer. The DRAM was invented by Robert Dennard at the IBM Thomas J. Watson Research Center in 1966. Global DRAM sales were more than $60 billion in 2020 according to the website MarketWatch. https://www.marketwatch.com/press-release/dynamic-random-access-memory-dram-market-size-in-2021-97-cagr-with-top-countr ies-data-business-growth-factors-share-industry-analysis-by-top-manufactu res-insights-and-forecasts-to-2026-2021-08-09.
18. Takada K, Yamagishi H, Minami H, Imai M (1998) In: Huff H, Tsuya H, Gösele U (eds) Semiconductor Silicon/1998. The Electrochemical Society, Pennington, p. 376.
19. Bridgman PW Crystals and their Manufacture US 1,793,672 Filed February 16, 1926; issued February 24, 1931. His crystal growth method was important in launching the early work on the transistor. Bridgman was awarded the 1946 Nobel Prize in Physics for his work in high-pressure physics,
20. Al-Hamdi TK, McPherson SW, Swain SK, Jennings J, Duenow JN, et al. (2020) CdTe synthesis and crystal growth using the high-pressure Bridgman technique. Journal of Crystal Growth 534:125466.
21. Shimura F (2017) Single-crystal silicon: growth and properties. In: Kasap S, Capper P. (eds) Springer Handbook of Electronic and Photonic Materials. Springer Handbooks. Springer, Cham. All materials contain point defects, such as vacancies, where there is a missing atom in one of the positions in the crystal structure.
22. Kane BE (1998) A silicon-based nuclear spin quantum computer. Nature 393:133.
23. Itoh KM (2005) An all-silicon linear chain NMR quantum computer. Solid State Communications 133:747–752.

24. Kao KC, Hockham GA (1966). Dielectric-fibre surface waveguides for optical frequencies. Proceedings of the Institution of Electrical Engineers 113:1151–1158. The classic paper by Kao and Hockham that launched optical fiber communication.

25. Charles K. Kao—Facts. NobelPrize.org. Nobel Prize Outreach AB 2021. Tue. 19 Oct 2021. https://www.nobelprize.org/prizes/physics/2009/kao/facts/.

26. MacChesney JB, O'Connor PB, DiMarcello FV, Simpson JR, Lazay PD (1974) Preparation of low loss optical fibers using simultaneous vapor deposition and fusion. In: Proceedings 10th International Congress on Glass pp. 6–40-6–44.

27. Nagel SR, MacChesney JB, Walker KL (1982) An overview of the modified chemical vapor deposition (MCVD) process and performance. IEEE Transactions on Microwave Theory and Techniques MTT-30:305–322.

28. Quote reproduced in lecture given by Mrs. Gwen MW Kao on behalf of Professor Charles K. Kao on December 8, 2009 at Stockholm University.

29. Staudinger H, Frost W (1935) Ber. Dart. Chem. Ges., 68:2351.

30. Hermann Staudinger—Biographical. NobelPrize.org. Nobel Prize Outreach AB 2021. Tue. 19 Oct 2021. https://www.nobelprize.org/prizes/chemistry/1953/staudinger/biographical/.

31. Demirors M (2011) The history of polyethylene. In: 100 + Years of Plastic. Leo Baekeland and Beyond, ACS Symposium Series, American Chemical Society, Washington pp 115–145.

32. Wilson GD (1994) Polythene: The early years. In: Mossman STI, Morris PJT (eds) The Development of Plastics. The Royal Society of Chemistry, Cambridge. Imperial Chemical Industries (ICI) based in the United Kingdom developed polyethylene (polythene as it is known in the UK) during the period 1933–1935. In the early 1950s ICI licensed the technology to other companies.

33. The average distance between the Earth and Mars is about 140 million miles. Approximately every two years a Mars Close Approach occurs where the two planets are at their smallest separation. The next Mars Close Approach is December 8, 2022, when the planets will be only a mere 38.6 million miles apart.

34. Honniball CI, Lucey PG, Li S, Shenoy S, Orlando TM, Hibbitts CA, Hurley DM, Farrell WM (2020) Molecular water detected on the sunlit Moon by SOFIA Nature Astronomy 5:121–127.

35. Morries RV, Score R, Dardano C, Heiken G (1983) Handbook of Lunar Soils Lyndon B. Johnson Space Center, Houston TX.

36. Clarke FW, Washington HS (1924) The composition of the Earth's crust Professional Paper 127 Department of the Interior, Washington DC. Available online https://pubs.usgs.gov/pp/0127/report.pdf. Accessed 10 August 2022.

37. Goulas A, Binner JGP, Harris RA, Friel RJ (2017) Assessing extraterrestrial regolith material simulants for in-situ resource utilisation based 3D printing Applied Materials Today 6:54–61.
38. Postiglione G, Natale G, Griffini G, Levi M, Turri S (2015) Conductive 3D microstructures by direct 3D printing of polymer/carbon nanocomposites via liquid deposition modeling. Composites: Part A 76:110–114.

5

There's *Still* Plenty of Room at the Bottom

When California Institute of Technology physicist Richard Feynman was describing how to write the entire 24 volumes of the *Encyclopedia Britannica*, all 44 million words of it, on the head of a pin by manipulating clusters of individual atoms to form tiny letters he was describing what would eventually be known as an example of "*nano*technology'—technology conducted at the nanometer scale.[1] Each dot that collectively makes up the words on a printed page when reduced to fit on the head of a pin would be only about 30 atoms across (on the order of 10 nm or less). Moving these atomic clusters to form microscopically small letters would truly be technology at the *nano*scale.

In December 1959 at the American Physical Society meeting held in Pasadena, California Richard Feynman, who had already achieved fame for his eponymous diagrams that illustrated the interactions between fundamental particles such as between two electrons, explored the possibilities of miniaturization in a talk entitled "There's Plenty of Room at the Bottom". This presentation like much of the topics Feynman wrote or spoke about during an amazing career was both visionary and inspiring. This particular lecture is often considered to mark the dawn of the field of nanotechnology, even though the term was not used by Feynman, nor anyone else, until 1974.

The term "Nanotechnology" was coined fifteen years after Feynman's talk in a 1974 presentation by Norio Taniguchi, a professor at Tokyo Science University in Japan.[2] Professor Taniguchi's paper was describing trends in materials processing at length scales of 100 nm or less (the typical range that is considered to be "nano"). A particular challenge Taniguchi identified was

© The Author(s), under exclusive license to Springer Nature
Switzerland AG 2023
M. G. Norton, *A Modern History of Materials*,
https://doi.org/10.1007/978-3-031-23990-8_5

how to produce semiconductor devices with feature sizes down to 100 nm, possibly even as small as 50 nm – in the realm of so-called large-scale integration or LSI. The necessary dimensions for enabling LSI were reached by 2005, three decades after Taniguchi's presentation. Feature sizes on current integrated circuits are now as small as 14 nm with research attempting at shrinking that distance to only 7 nm (Significantly smaller than the size of a coronavirus particle!).

To place Feynman's presentation in the context of the state of technology at the end of the nineteen fifties, it was only one year before his talk in Pasadena that Texas Instruments' Jack Kilby had constructed and tested a very basic integrated circuit made using a germanium crystal with the circuit components connected by "flying" gold wires. The first planar integrated circuit, the precursor of the modern silicon chip, was not built until May 1960, almost six months after Feynman's lecture. In 1959 computers were far from being miniaturized. They were, as Feynman noted, "too big" and that to do large amounts of computation such as that needed for facial recognition would require a computer "the size of the Pentagon." Today, advances in nanotechnology have enabled a handheld computer, aka a "smartphone" to have millions of times more power than the banks of computers that allowed three Apollo 11 astronauts to head to the Moon and return safely to Earth and to provide facial recognition, for instance Apple's Face ID, all while fitting snuggly in one hand.[3]

Although electron microscopes existed, and were commercially available, in 1959 they were not able to achieve atomic resolution—they needed to be improved by a factor of 100. But Feynman, ever the visionary, explained how they might in the future have a resolving power small enough that they could be used to read the tiny letters formed from atomic clusters on the head of a pin. But before we could observe these structures it would be necessary to make them. What would represent the ultimate in manipulation would be the ability to arrange individual atoms in any which way we want—one atom at a time. This challenge Feynman considered to be something that might be possible "in the great future."

But that great future was closer than even the great optimist Richard Feynman might have expected.

On September 28, 1989, Don Eigler a scientist at IBM's Almaden Research Center in California moved a single xenon atom back and forth across a platinum surface. Eigler realized the importance of his experiment writing in his notebook, "first atom to be manipulated under control." Later that same year, Eigler and Erhard Schweizer, a visiting scientist from the Fritz Haber Institute in Berlin cooled down a nickel crystal to a temperature of just

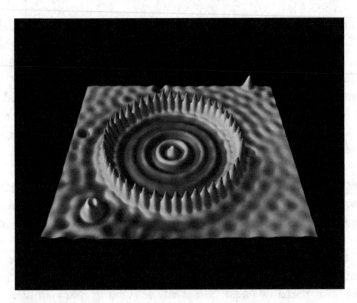

Fig. 5.1 Scanning tunneling microscope image of a structure known as a quantum corral. Each of the sharp multi-colored peaks is a single iron atom. Forty eight of them were moved to create a circle (Image obtained by Don Eigler, IBM Almaden Research Center. Reprinted with permission of the NISE Network)

4 K (four degrees above the absolute zero of temperature; a frigid -269°C) and using the tip of a scanning tunneling microscope they dragged one at a time over a period of 22 h 35 xenon atoms to spell out the three letters "I B M".[4]

In 1993 Eigler and IBM colleagues Michael Crommie and Christopher Lutz constructed an even more ambitious structure known as a quantum corral, which is shown in Fig. 5.1, consisting of 48 iron atoms. Each iron atom was moved across a copper surface using the tip of a low-temperature scanning tunneling microscope, similar to that previously used to move the xenon atoms, creating a circle that is about 14 nm across.[5] The circular wave patterns inside the corral are formed by copper electrons trapped by the enclosing ring of iron atoms. Although there were no immediate uses identified for the quantum corral its importance is in illustrating the possibility of how to precisely manipulate individual atoms to create novel structures with dimensions at the nanoscale.

The potential of nanotechnology was not lost on the United States government. On December 3, 2003, President George W. Bush signed into law the twenty-first century Nanotechnology Research and Development Act or NRDA, which led to the formation of the National Nanotechnology Initiative (NNI). The NNI created a broad definition of nanotechnology as:

" … science, engineering, and technology conducted at the nanoscale (1 to 100 nm), where unique phenomena enable novel applications in a wide range of fields, from chemistry physics and biology, to medicine, engineering and electronics."[6]

Accompanying the signing of the Nanotechnology Research and Development Act the White House press release highlighted the evolutionary and revolutionary promise of nanotechnology—improving and creating entirely new products and processes in areas from electronics to healthcare:

- Carbon nanotubes are essentially sheets of graphite rolled into extremely narrow tubes—a few nanometers in diameter. Because of their nanoscale size and excellent conductivity, carbon nanotubes are being studied as the possible building blocks of future electronic devices.
- Nanotechnology may one day enable the detection of disease on the cellular level and the targeting of treatment only to tissues where it is needed in a patient's body, potentially alleviating many unpleasant and sometimes harmful side effects.
- Nano-manufacturing of parts and materials "from the bottom up"—by assembling them on an atom-by-atom basis—may one day be used to reduce waste and pollution in the manufacturing process.
- Nanosensors already are being developed to allow fast, reliable, real-time monitoring for everything from chemical attack to environmental leaks.

Nanotechnology's potential to provide cleaner energy was also noted:

- Woven into a cable, carbon nanotubes could provide electricity transmission lines with substantially improved performance over current power lines.
- Certain nanomaterials show promise for use in making more efficient solar cells and the next-generation catalysts and membranes that will be used in hydrogen-powered fuel cells.

The Nanotechnology Research and Development Act itself did not call out any individual material or materials but rather described establishing programs, setting up panels to explore the potential of nanotechnology, and providing the necessary funding to make progress in this exciting area. However, the White House press release did highlight one material in particular—the carbon nanotube. When the Nanotechnology Research and Development Act was written into law carbon nanotubes were the cause of a great deal of excitement within the scientific community because of their apparently unlimited potential in everything from overhead power cables to ultrarapid chemical sensors.

Carbon nanotubes were first observed in 1991 by Japanese scientist Sumio Iijima working at the NEC Corporation in Tsukuba, Japan.[7] Using a high-resolution electron microscope housed in the Fundamental Research Laboratories of NEC, Iijima was examining carbon samples made by the technique of arc-discharge evaporation. Using a DC electric field a carbon soot was vaporized and the resulting product that collected at various locations inside the reaction chamber was removed. This was a similar method that had been used to produce large quantities of the spherical C_{60} molecules that became known as buckyballs after the geodesic domes of American architect Richard Buckminster Fuller and their similarity to a soccer ball.[8]

When the samples that Iijima had collected from the reactor were placed into the electron microscope a number of long thin cylindrical structures were observed that he interchangeably referred to as "needles" or "microtubules." Three examples of these needle-like structures are shown in Fig. 5.2. The dark lines running from top to bottom of the images are planes of carbon atoms. In cross section the structures resemble onions or Russian dolls—one carbon layer snuggly wrapped around the one underneath. In some cases, the number of layers were as few as two. In other cases, there may be as many as fifty layers. Carefully analyzing the structures, it was clear to Iijima that each layer was a single sheet of carbon atoms that had been rolled up into a tube—a nanotube. Carbon nanotubes with one layer of carbon atoms are known as single-walled as opposed to the multi-walled variety that have many layers. Sumio Iijima and Toshinari Ichihashi went on to discover single-walled carbon nanotubes in 1993, two years after they had first identified multi-walled carbon nanotubes.[9]

Although Iijima's groundbreaking 1991 *Nature* paper didn't report any measured properties of carbon nanotubes nor did it mention any potential applications for these unusual structures it was not long before others did.

Carbon nanotubes were found to have an extraordinary ability to conduct both electricity and heat. The electrical conductivity of single-walled – one layer thick – carbon nanotubes is comparable to that of the most conductive metals silver and copper. In terms of thermal conductivity, carbon nanotubes are over fifteen times more conductive than copper and comparable to diamond, the material with the highest thermal conductivity of all.[10] Mechanically, carbon nanotubes have tensile strengths exceeding those of steel and Kevlar (the bullet-proof vest material). The strength comes from the extremely strong bonding between carbon atoms, which is greater than even

Fig. 5.2 The first electron microscope images of carbon nanotubes. A cross-section of each tube is illustrated below. The tube in image *a* consists of five layers of graphite, *b* is a two-sheet tube, while *c* is a seven-sheet tube. Although the images might not appear impressive their impact on the field of nanotechnology was highly significant (Reprinted from Iijima S (1991) Helical microtubules of graphitic carbon. Nature 354:56–58 with permission from Springer Nature)

that found in diamond.[11] And when bent, carbon nanotubes rapidly spring back to their original shape demonstrating excellent elasticity.

These unique properties sparked an extraordinary amount of carbon nanotube research within the scientific community both in industrial laboratories and at universities. It also led to many proposed, even fanciful, applications for these intriguing and novel structures. From next generation transistors to tiny wires for circuit interconnects, to tear-resistant fabrics, and even using carbon nanotube cables for a space elevator the number of potential applications seemed to be only limited by the imagination.

All the excitement generated by the discovery of carbon nanotubes was matched with an exponential growth in the number of papers written on nanoscience and nanotechnology, particularly papers featuring carbon nanotubes.[12] In practical terms, the discovery of carbon nanotubes was perhaps a more significant beginning to the field of nanotechnology than Feynman's visionary 1959 presentation to an assembled group of physicists in Pasadena. Carbon nanotubes were relatively easy to produce, certainly when compared to the technical challenges of moving individual atoms around on a surface to produce a nanostructure. Synthesis techniques were straightforward requiring equipment that was widely available in laboratories around the world. Scientists were not just limited to the arc-discharge evaporation method used by Sumio Iijima. A whole range of additional techniques including laser ablation and chemical vapor deposition were found to produce usable quantities of carbon nanotubes for research purposes.

Scientists quickly realized that there might be other nanostructured forms of carbon that could be synthesized in the laboratory and these too might have some very interesting and potentially unique properties.

Unwrapping a single-walled carbon nanotube produces a one-atom-thick layer of carbon called graphene. The excitement around graphene, once it had been successfully isolated, was equal to that created by the discovery of carbon nanotubes a little more than a decade earlier.

The 2010 Nobel Prize in Physics was awarded to University of Manchester physicists Andre Geim and Konstantin Novoselov for " … groundbreaking experiments regarding the two-dimensional material graphene."[13] In 2004 Geim and Novoselov produced the single sheet of carbon atoms using nothing more than a piece of graphite and some adhesive "Scotch" tape.[14] By repeatedly removing layer after layer of carbon atoms the two Russian-born scientists were eventually left with just a single layer of carbon atoms attached to the tape. Unlike carbon nanotubes, which are considered a one-dimensional nanomaterial—they are only a few nanometers wide but can be many tens of nanometers, or more, long—graphene is a two-dimensional nanomaterial. Atomically thin in one dimension, graphene sheets can be up to several fractions of a millimeter wide.[15,16]

The importance of isolating graphene and why the work was so rapidly recognized by the Royal Swedish Academy of Sciences was two-fold. Firstly, the material was completely new. There was considerable doubt whether such a thin crystalline material—just one atom thick—would even be stable. Secondly, graphene quickly demonstrated many outstanding properties from its high strength to excellent thermal and electrical conductivities. Not only is graphene so thin it is almost completely transparent to visible light it is so dense that even the tiny helium atom at just 0.098 nm across cannot pass through.

Until 1985 carbon was considered to exist in two natural allotropes (different crystalline forms of the same chemical element). These were diamond and graphite. Even though diamond is not the thermodynamically stable form of carbon at room temperature its transformation to graphite is so slow that both diamond and graphite co-exist in nature. What is significant about the unique electron arrangements in graphite (sp^2) and diamond (sp^3) is that they lead to the two materials having very distinctive properties. We can most easily see this difference by comparing a black piece of graphite with the highly transparent reflective sparkle of a well-cut diamond. Whereas diamond is an excellent electrical insulator, graphite is a good conductor of electricity. Mechanically, the two materials are also very different. Diamond is very hard, the hardest of all known minerals, while graphite can be easily sheared breaking the weak bonds between adjacent layers of carbon atoms. Diamond is invaluable in industry because it can be used to cut and polish any material. No material is harder. On the other hand, graphite makes a good "lead' for pencils leaving a soft trail of carbon on a piece of paper. For the same reason it is also an excellent lubricant.

If a single layer of carbon atoms is sheared from a piece of graphite, then we have formed graphene. It was this innovation that led to the recognition of Andre Geim and Konstantin Novoselov by the Royal Swedish Academy of Sciences.

Graphene is one of the newly identified allotropic forms of carbon, which exist in nature but have required a certain amount of serendipity in their synthesis and isolation. Identifying these allotropes has required the use of advanced instrumentation such as high-resolution electron microscopes. The building block of all of these various carbon structures—except for diamond—is the graphene sheet as illustrated in Fig. 5.3. When the single sheet of carbon atoms is rolled into a ball we produce the spherical fullerene molecule consisting of sixty carbon atoms arranged in hexagons and pentagons resembling the shape of a soccer ball. When wrapped into a tube a graphene sheet becomes a carbon nanotube. When stacked one on top of the other the graphene sheets become graphite.

The first of these elusive carbon allotropes was identified in 1985 by Harry Kroto, Robert Curl and Richard Smalley while all three were at Rice University in Houston, Texas. They went on to share the 1996 Nobel Prize in Chemistry for " … their discovery of fullerenes."[17] The background to the experimental work that led to this important discovery was to understand how long-chain carbon molecules formed in interstellar space.[18] Using a neodymium-YAG laser, a type of laser used in some surgical procedures, a sample of graphite was vaporized and the carbon species that formed were

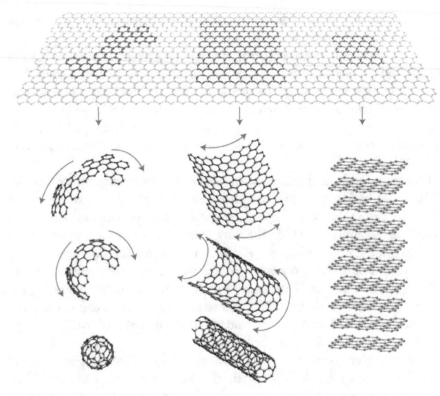

Fig. 5.3 Mother of all graphitic forms. Graphene is a two-dimensional (2D) building material for carbon materials of all other dimensionalities. It can be wrapped up into zero-dimensional (0D) buckyballs, rolled into one-dimensional (1D) nanotubes or stacked into three-dimensional (3D) graphite. The importance of the layered structure of graphite and the spacing between layers is discussed further in Chapter 6 (Reprinted from Geim AK, Novoselov KS (2007) The rise of graphene. Nature Materials 6:184 with permission from Springer Nature)

collected. These species were subsequently analyzed using a technique called mass spectroscopy. The results suggested that one of the structures was a molecule of sixty carbon atoms, C_{60}, arranged in the form of a truncated icosahedron resembling a soccer ball.

Kroto and his colleagues speculated on the importance of their discovery including the possibility of derivative species such as $C_{60}F_{60}$ that because of the presence of fluorine might be a super lubricant demonstrating the ultra-slippery non-stick properties exhibited by the fluorocarbon polymer Teflon. They also speculated that a buckyball might act as a cage that could be filled with another type of atom, for instance, lanthanum or oxygen, which might create some unusual properties leading to new forms of catalyst. Catalysts play an essential role in industry. Over 80% of all manufactured products and approximately 90% of all industrial chemicals produced in the world use catalysts during the manufacturing process.

But an over-riding requirement that would be necessary to realize the full chemical and practical value of C_{60} was "a large-scale synthetic route." In those early days immediately following the initial experiments it was challenging to obtain just enough material for a more detailed analysis and further characterization. But the situation changed in 1990 with the publication in the journal *Nature* of a process that was able to produce significant quantities of buckyballs.

Even with the development of methods to produce macroscopic quantities of fullerenes leading to a more complete understanding of their structure and properties the applications for these unusual molecules remained elusive. At a lunch in Washington D.C in the early 2000's I was fortunate to sit next to Sir Harry and his wife Margaret. Kroto lamented that his discovery had not led to groundbreaking applications in technology and commented that the only application that had been found was in cosmetics made by French giant L'Oreal. Skin care products such as make-up, sunscreen, and hair conditioner make use of the anti-oxidative behavior of fullerenes coupled with their light absorbing and reflecting properties.[19]

In the original *Science* paper describing the formation of graphene the authors provided a detailed characterization of their novel material and identified one possible application of graphene, next generation beyond-silicon transistors. These graphene-based transistors would be even smaller than current technology, consume less energy, and operate at higher frequencies. All told, graphene transistors would result in more efficient computers.

A team in Aachen, Germany at the Advanced Microelectronic Center and the Rheinisch-Westfälische Technische Hochschule (RWTH) fabricated the first graphene transistor in 2007.[20] An image of the device recorded using a scanning electron microscope is shown in Fig. 5.4. This type of transistor known as a field effect transistor or FET is the most important geometry in fabricating computer chips. All field effect transistors have three components a source (shown as S in the image), a drain (D) and a gate (G). The voltage applied to the gate determines whether an electrical current will flow between the source and the drain. In the image the total length from source to drain is 7.3 μm. At the gate the graphene strip is only 265 nm wide with a gate length of 500 nm. A field effect transistor works by controlling the electric current that flows between the source and drain by varying the voltage applied to the gate.

Despite graphene's many remarkable properties that make it of interest as a rival to silicon for next-generation transistors and the fact that the transistor

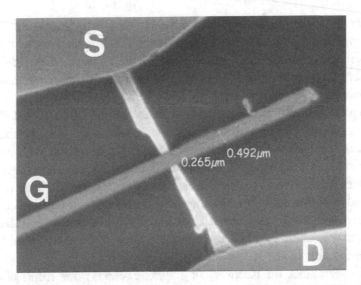

Fig. 5.4 Scanning electron microscope image of a graphene transistor. The graphene strip is the lighter contrast region between the source (S) and the drain (D). The transistor was formed on an oxidized silicon wafer (Reprinted with permission from Lemme MC, Echtermeyer TJ, Baus M, Kurz H (2007) A graphene field effect device. IEEE Electron Device Lett 28:10. Copyright 2007 IEEE)

effect has been demonstrated in graphene, considerable challenges remain before we will see graphene transistors controlling our smartphones and other electronic devices. Included amongst these challenges is the significant limitation of producing large high-quality graphene sheets using the current range of synthesis methods. Certainly, it is impossible to consider that using Scotch tape to peel layer after layer of graphene from a graphite block is an industrially scalable method. Even some of the alternative methods being used in university laboratories are expensive and difficult to scale to mass production type volumes. Less costly techniques tend to produce graphene layers of uneven quality and thickness.

A recent review of graphene electronics by Mohamed Warda in the Department of Physics at Imperial College London makes a clear case of the uphill challenge faced by graphene transistors, " … it is unclear whether graphene will ever replace silicon in modern consumer electronics at large, for the … reasons regarding the difficulty of … [synthesizing] high mobility graphene samples on a large scale. Nevertheless, it is becoming more apparent that graphene could play an important role in more specialized areas of modern electronic engineering, such as terahertz technology."[21]

So, while graphene transistors were fabricated within a few years after the discovery of the material their impact may be limited by the continued advances in silicon technology and developments in quantum computing. Geim and Novoselov recognized that transistors might not be the most impactful use of graphene. They concluded their *Science* paper by stating that "nontransistor applications of this atomically thin material ultimately may prove to be the most exciting."

The press release accompanying the award of the 2010 Nobel Prize in Physics to Geim and Novoselov noted the combination of graphene's high transparency together with its excellent electrical conductivity with proposed applications in transparent touch screens, light panels, and maybe even solar cells. Over a dozen years later none of these applications have been commercially realized. But scientists working in this area remain optimistic. At the Graphene Institute at the University of Manchester where the wonder material was first isolated they believe that the impact of graphene could be on a scale comparable to the Industrial Revolution.

In the years since the discovery of graphene several other forms of carbon have been predicted based on computational studies.[22] For instance, one of these novel allotropes is known as T-carbon with a diamond-like structure, but each carbon atom in the unit cell is replaced by a tiny tetrahedron consisting of four carbon atoms (C_4). An interesting and potentially very useful property of T-carbon is its very low density—less than half that of diamond—with a large space between carbon atoms that might be used, for instance, to store hydrogen. Safe and compact storage of hydrogen is a key enabling technology for the advancement of hydrogen and fuel cell technologies for a whole host of stationary, portable and transportation applications.

Working at the nanoscale by manipulating individual atoms is a relatively recent development. Writing the three letters I, B, and M using xenon atoms happened less than 40 years ago. The term that we use to define everything from synthesizing nanomaterials through to using them in a wide range of applications was coined in 1974. However, exploiting the unique properties of nanomaterials is not new. One of the most striking early examples of the use of nanomaterials was mentioned in Chapter 3, the fourth century Roman drinking cup—the Lycurgus cup—on display in the British Museum in London. What makes the Lycurgus cup of such historical importance and scientific interest is that it is dichroic—it has a different color when viewed using transmitted light than it does in reflected light.

The origin of the cup's color was revealed when small samples of the glass were examined using a transmission electron microscope.[23] The glass was found to contain metallic particles of both gold and silver that were between 50 and 100 nm in diameter, well within the range of what we define as

Fig. 5.5 Example of a glaze color produced by copper nanoparticles (Image by sailko reprinted under the Creative Commons Attribution-ShareAlike 3.0 license)

nanoparticles. Exactly how the Roman's made this rare and unusual glass of which there are very few remaining examples is unknown and will probably remain a mystery. Possibly like several early materials discoveries it happened by a fortunate accident.

Nanoparticles of silver, copper, and other metals were widely used from the ninth century onwards for creating color in ceramic glazes. The example shown in Fig. 5.5 is a fifteenth century vase from the Italian city of Deruta. The amber color in this striking example of Renaissance pottery is due to the presence of copper nanoparticles. Gold and silver nanoparticles also produced the luminous red and yellow colors seen in medieval stain glass windows produced in Europe beginning in the twelfth century. Unfortunately, only a small proportion of these earliest windows have survived.

With their unique optical and electronic properties gold nanoparticles are the most interesting and possibly the most useful of all nanomaterials. Their unusual properties were recognized by Michael Faraday and described, in great detail, by him in his Bakerian Prize Lecture delivered at the Royal Society in London on February 5, 1857.[24] Faraday was the Fullerian Professor of Chemistry at the Royal Institution in London; the first person to hold that lifetime professorship. At the time of his lecture, Michael Faraday was already very well known for his work in electrolysis (Faraday's Laws), the magneto-optical properties of materials (the Faraday effect), static electricity (the Faraday cage), among a great many other discoveries and inventions. Although Faraday didn't use the term "nanoparticle" or "nanomaterial" in his lecture he was, for the first time, describing how "finely-divided"—in other words nanosize—gold particles behaved differently from the bulk metal. This phenomenon would become known as the quantum size effect: how size affects properties at the nanoscale. Faraday's Royal Society lecture, rather than Feynman's American Physical Society lecture, which was delivered over 100 years later, is possibly an alternative beginning of our exploration of nanoscience and nanotechnology.

Faraday's lecture concerned how gold and other metals, including platinum, tin, and copper, interact with light. A considerable part of the lecture was devoted to Faraday's extensive experimentation using very thin sheets of gold leaf estimated to be only 1/282000th of an inch thick (about 90 nm). But Faraday also worked with gold wires and most importantly in terms of the history of nanotechnology with tiny gold particles that were, described as being, considerably smaller than the wavelength of light.

Visible light ranges from short wavelength violet at 380 nm to red at the much longer 740 nm. Faraday was correct in his assumption that " … the waves of light are so large compared to the dimensions of the particles of gold." In fact, they were likely to be ten or maybe as much as 100 times smaller than the wavelength of light. Faraday went on to state, " … it is clear that that metal, reduced to small dimensions by mere mechanical means, can appear of two colours (*sic*) by transmitted light, whatever the cause of the difference may be. The occurrence of these two states may prepare one's mind for the other differences with respect to colour (*sic*), and the action of the metallic particles on light. Which have yet to be described."

Faraday noted that larger gold particles—those that settled very quickly through the solution—produced a blue color. A ruby red color was produced when the gold particles were of a much smaller size. This observation apparently caused Faraday some concern leading him to state, "But that the blue particles are always merely larger particles does not seem admissible for a moment …; there is probably some physical change in the condition of the particles, caused by the presence of the salt and such affecting media, which is not a change of the gold as gold, but rather a change of the relation of the surface of the particles to the surrounding medium."

But Faraday's concerns turned out to be unnecessary. The interaction of gold nanoparticles with light is strongly dictated by their size. For small (~30 nm) gold nanoparticles light is absorbed in the blue-green portion of the electromagnetic spectrum (~450 nm) while red light (~700 nm) is reflected, yielding a rich red color. As the particle size increases, the absorption shifts to longer, redder wavelengths. Red light is absorbed and blue light is reflected yielding solutions with a pale blue or purple color. Faraday's experiments led him to the connection between the bright coloration of the purple of Cassius and the presence in the glass of very small gold particles.[25] This was over 130 years before the gold particles in the glass were seen using a powerful electron microscope.

Michael Faraday's observations that the color produced by "finely divided" gold depended on their size is just one example of what is called the quantum size effect. That the properties of nanomaterials are a function of their size. One hundred and twenty five years after Faraday's presentation to the Royal Society of London a paper published by Matsutake Haruta, a professor in the Department of Applied Chemistry and the Graduate School of Urban Environmental Sciences at Tokyo Metropolitan University in Japan reported quantum size effects in gold nanoparticle catalysts. Haruta demonstrated that gold, the least reactive of our metals and prized for its inertness, when in the form of nanoparticles 5 nm in diameter or less has extraordinary catalytic properties.[26,27] In fact, it is an excellent catalyst for a number of important reactions, in particular the oxidation of carbon monoxide to carbon dioxide. This reaction is fundamental to the operation of an automobile's catalytic converter. But what was most surprising was that not only was gold more active than the platinum catalyst the researchers compared it to but it worked at lower temperatures. This result is particularly important from an environmental point of view. Considerable amounts of pollutants are emitted from automobile tail pipes within the first five minutes after starting the engine. Current exhaust catalysts, based on alloys of platinum and rhodium, must be at a temperature of at least 350°C to function. So for the first part of a commute, we are spewing—unless we drive an electric car—raw emissions containing a potent mixture of unburnt fuel, carbon monoxide, and nitrogen dioxide into the atmosphere. These so-called cold-start emissions impose a major pollution problem requiring new catalysts that can operate at lower temperatures than the platinum rhodium alloys. Nanoparticle gold might be just that catalyst.

According to data from Grand View Research the global gold nanoparticle market is expected to reach $6.3 billion by 2025. What is principally driving this growth is applications in the healthcare industry. Gold nanoparticles can be used for both detecting and treating disease. Spherical gold nanoparticles, typically between 10 and 50 nm in diameter are at the heart of the hundreds of millions of Rapid Diagnostic Tests (RDTs) that are used around the world every year. As an example, malaria RDTs work by applying a single drop of the patient's blood to a test strip. Malaria antigens interact with the gold nanoparticles causing a color change on the test strip if the disease is present. The tests are simple, quick, reliable and robust. Furthermore they can be used without the need for expensive equipment or highly qualified personnel. According to the World Health Organization, over 300 million malaria RDTs are sold each year.[28]

Recently we have become all too familiar with the rapid diagnostic lateral flow tests used to diagnose COVID-19. These too rely on the power of gold nanoparticles to indicate the presence of the disease using a small saliva sample from the patient. A single red stripe indicates a negative COVID test, whereas two red stripes mean the patient is positive for COVID-19. There are now at least 400 rapid diagnostic tests commercially available worldwide.[29]

New diagnostic technologies exploiting the unique properties of gold nanoparticles are continually being developed. A team at Imperial College in London has demonstrated a simple diagnostic test for HIV that allows the disease to be spotted earlier and does not require the use of expensive instrumentation.[30] Their approach uses gold nanoparticles, which are formed when hydrogen peroxide reacts with a solution containing gold ions. If the target molecule, an enzyme called catalase, is present it can break down the hydrogen peroxide. The speed of the reaction determines the shape of the resulting gold nanoparticles. If the growth is rapid, the gold nanoparticles are spherical causing the solution to turn red. When the nanoparticles grow slowly, they have more irregular shapes and the solution turns blue. The color change is obvious enough that it can be seen with the naked eye.

A similar approach has been used as an early detection for prostate cancer, the second-most common cancer among men in the United States. The high death toll is partly because the disease has few symptoms in its early stages, meaning that it is difficult to detect. Gold nanoparticles may provide a critical breakthrough in diagnosing this disease.

In addition to detecting disease, gold nanoparticles can be used to treat disease. One example is an injectable compound containing gold that was found to be effective in treating rheumatoid arthritis after the Second World War. In the twenty-first century, gold nanoparticles are being used extensively in the development of innovative medicines and medical techniques, including needleless vaccine delivery and antimicrobial agents. But perhaps the most exciting area of research is in the treatment of cancer.

According to the World Health Organization, there were nearly 10 million cancer deaths in 2020. This is despite advances in diagnosis and treatment. Treating cancer is made particularly difficult because often the drugs used in chemotherapy can damage healthy cells. Treatments that target cancer cells directly while limiting the impact on the rest of the body could increase the chance of a full recovery. Research in gold nanoparticles is leading to targeted techniques of delivering drugs and other cancer treatments. With support from the National Cancer Institute CytImmune, an American-based pharmaceutical company, has developed a method of delivering anti-cancer drugs directly to tumors using gold nanoparticles.[31] The drug is bound to the gold nanoparticles, which are injected into the bloodstream and travel to the site of the tumor, treating it while leaving surrounding tissue largely unaffected.

Another company, Nanospectra Biosciences, is taking a different approach to using gold nanoparticles to treat cancer. In particular, they are focused on cancerous tumors within the prostate. The company has created "nanoshells," consisting of a gold-coated silica nanoparticle.[32] When these tiny particles are injected into the tumor and then irradiated with a laser of the appropriate frequency they heat up destroying the cancer cells in a process called thermal ablation.

Since Richard Feynman's seminal 1959 presentation to an assembled group of physicists in Pasadena many of the things he imagined have happened— we can see individual atoms, we can move them around on a surface to create tiny structures, and we have computers so small that they fit within the palm of a hand. We also have a name for all of this, it is "nanotechnology". Nanotechnology has been promoted as a technology that will revolutionize the twenty-first century. We are already seeing its enormous impact particularly in medicine, but there is still plenty of room at the bottom.

Notes

1. Feynman RP (1959) Plenty of room at the bottom. Transcript of a talk presented to the American Physical Society in Pasadena on December 1959. Available online at: https://web.pa.msu.edu/people/yang/RFeynman_plentySpace.pdf. Accessed 10 Aug 2022.
2. Taniguchi N (1974) On the basic concept of 'nano-technology. Proceedings of the International Conference on Production Engineering, Tokyo, Part II, Japan Society of Precision Engineering, Tokyo, pp 18–23. Taniguchi N (1983) Current status in, and future trends of, ultraprecision machining and ultrafine materials processing. CIRP Annals 32:573–582. A review paper that describes trends towards achieving machining and processing at the nanoscale.
3. About Face ID advanced technology. https://support.apple.com/en-us/HT208108. Accessed 10 Aug 2022.
4. Eigler DM, Schweizer EK (1990) Positioning single atoms with a scanning tunnelling microscope. Nature 344:524–526.
5. Crommie MF, Lutz CP, Eigler D (1993) Confinement of electrons to quantum corrals on a metal surface. Science 262:218–220.
6. National Nanotechnology Initiative (NNI); Available online: www.nano.gov. Accessed 10 Aug 2022.
7. Iijima S (1991) Helical microtubules of graphitic carbon. Nature 354:56–58. This paper has been cited over 55,000 times.
8. Krätschmer W, Lamb LD, Fostiropoulos K, Huffman, DR (1990) Solid C_{60}: a new form of carbon. Nature 347:354–358.
9. Iijima S, Ichihashi T (1993) Single-shell carbon nanotubes of 1-nm diameter. Nature 363:603–605. This paper was received 23 April 1993 and accepted on 1 June 1993. The paper immediately following Iijima's and Ichihashi's in volume 363 was by D.S. Bethune and colleagues at IBM Almaden Research Center. Their paper was received 24 May 1993 and accepted 3 June 1993.
10. Berber S, Kwon Y-K, Tománek D (2000) Unusually high thermal conductivity of carbon nanotubes. Physical Review Letters 84:4613.
11. The strength of carbon nanotubes is due to the sp^2 bonding between carbon atoms. This is stronger than the sp^3 bonding that occurs in diamond.
12. Kwon S (2016) Bibliometric analysis of nanotechnology field development—from 1990 to 2015. Presentation to the Southeastern Nanotechnology Infrastructure Corridor (SENIC) meeting, Atlanta. There were about 2,000 papers published in 1991, the year of Sumio Iijima's paper announcing carbon nanotubes. Over 30,000 papers were published in 2002 and that number shot up to over 110,000 ten years later.
13. Prize announcement. NobelPrize.org. Nobel Media AB 2021. Mon. 8 Mar 2021. https://www.nobelprize.org/prizes/physics/2010/prize-announcement/.

14. Novoselov KS, Geim AK, Morozov SV, Jiang D, Zhang Y, Dubonos SV, Grigorieva IV and Firsov, AA (2004) Electric field effect in atomically thin carbon films. Science 306:666–669.

15. Lopes LC, da Silva LC, Vaz BG, Oliveira ARM, Oliveira MM, Rocco ML, Orth ES, Zarbin AJG (2018) Facile room temperature synthesis of large graphene sheets from simple molecules. Chemical Science 9:7289–7428.

16. Tung VC, Allen MJ, Yang Y, Kaner RB (2009) High-throughput solution processing of large-scale graphene. Nature Nanotechnology 4:25–29.

17. Sir Harold Kroto – Biographical. NobelPrize.org. Nobel Prize Outreach AB 2022. Fri. 11 Mar 2022. https://www.nobelprize.org/prizes/chemistry/1996/kroto/biographical/.

18. Kroto HW, Heath JR, O'Brien SC, Curl RF, Smalley RE (1985) C_{60}: Buckminsterfullerene. Nature 318:162–163.

19. Mellul M (1997) Cosmetic make-up composition containing a fullerene or mixture of fullerenes as a pigmenting agent US 5,612,021. Filed 30 May 1995; Date of Patent 18 March 1997. One of several patents filed by French cosmetics giant L'Oreal for use of fullerenes in make-up and skin care products.

20. Lemme MC, Echtermeyer TJ, Baus M, Kurz H (2007) A graphene field-effect device. IEEE Electron Device Letters 28:282–284.

21. Warda M (2020) Graphene field effect transistors. Available at https://arxiv.org/pdf/2010.10382.pdf. Accessed 10 Aug 2022.

22. Yang L, He HY, Pan BC (2013) Theoretical prediction of new carbon allotropes. Journal of Chemical Physics 138:024502.

23. Barber DJ, Freestone IC (1990) An investigation of the origin of the colour of the Lycurgus Cup by analytical transmission electron microscopy. Archaeometry 32:33–45.

24. Faraday M (1857) The Bakerian Lecture: Experimental relations of gold (and other metals) to light. Philosophical Transactions of the Royal Society of London 147:145–181.

25. The process of making ruby red glass, one of many technological accomplishments of the Romans, was lost in Europe for over one thousand years until it was rediscovered in Germany in the seventeenth century. The color is widely referred to as the "Purple of Cassius" after German physician Andreas Cassius (the younger) who has frequently been credited with its rediscovery in 1685 following the publication of his work *De Auro* in Hamburg.

26. Haruta M, Sano H (1983) Preparation of highly active composite oxides of silver for hydrogen and carbon monoxide oxidation. Studies in Surface Science and Catalysis 16:225–236.

27. Haruta M, Kobayashi T, Sano H, Yamada N (1987) Novel gold catalysts for the oxidation of carbon monoxide at a temperature far below 0 C. Chemistry Letters 16:405–408. This has been cited more than 3,500 times. Haruta's definitive review of 1997 (Size- and support- dependency in the catalysis of gold. Catalysis Today 36:153–66) has attracted more than 5,000 citations.

28. Incardona S, Serra-Casas E, Champouillon N, Nsanzabana C, Cunningham J, Gonzalez IJ (2017) Global survey of malaria rapid diagnostic test (RDT) sales, procurement and lot verification practices: assessing the use of the WHO–FIND Malaria RDT Evaluation Programme (2011–2014). Malaria Journal 16:196.

29. Drain PK. (2022) Rapid diagnostic testing for SARS-CoV-2. New England Journal of Medicine 386:264–272.

30. de la Rica R, Stephens MM (2012) Plasmonic ELISA for the ultrasensitive detection of disease biomarkers with the naked eye. Nature Nanotechnology 7:821–824.

31. Nilubol N, Yuan Z, Paciotti GF, Tamarkin L, Sanchez C, Gaskins K, Freedman EM, Cao S, Zhao J, Kingston DGI, Libutti SK, Kebebew, E (2018) Novel dual-action targeted nanomedicine in mice with metastatic thyroid cancer and pancreatic neuroendocrine tumors. Journal of the National Cancer Institute 110:1019.

32. Rastinehad AR, Anastos H, Wajswol E, Winoker JS, Sfakianos JP, Doppalapudi SK, Carrick MR, Knauer CJ, Taouli B, Lewis SC, Tewari AK, Schwartz JA, Canfield SE, George AK, West JL, Halas NJ (2019) Gold nanoshell-localized photothermal ablation of prostate tumors in a clinical pilot device study. Proceedings of the National Academy of Sciences 116:18590-18596.

6

The Future of Mobility

Nothing is currently more essential in enabling our mobile world than lithium-ion batteries. If you can carry it around and it requires a source of electricity, a fair bet is it contains at least one lithium-ion battery. These creations of human ingenuity—requiring a detailed understanding of how materials interact at the atomic scale—power the ubiquitous smart devices that allow us to communicate with each other wherever we are in the world. Lithium-ion batteries are irreplaceable partners living in our pockets, handbags, and backpacks, until the point when we need to charge them so that they can continue their essential work—converting a reversible chemical process into electrical energy.

Lithium-ion batteries offer many advantages over competing battery technologies, particularly for mobile applications. For instance, lithium-ion batteries have a high energy density, they're lightweight, they can cycle many thousands of times without requiring any maintenance, and charging is both quick and easy. Lithium-ion batteries already dominate, by a considerable margin, the field of portable electronics. They are poised to occupy an equally dominant position in the rapidly growing electric vehicle market because they are simply the best performing technology available to date (and into the foreseeable future.)

The technological, environmental, and economic importance of lithium-ion batteries and the science necessary to deliver this technology was not lost on the Royal Swedish Academy of Sciences. In 2019 the Nobel Prize in Chemistry was shared equally between three scientists "for the development of lithium-ion batteries."[1] Two of those laureates, John B. Goodenough and M. Stanley Whittingham were born in Europe; Goodenough in Jena, Germany, and Whittingham in Nottingham, England. By the time of the award both had been resident in the United States for a considerable part of their lives. Whittingham's contribution was to devise and construct the first rechargeable lithium-ion battery. Goodenough performed groundbreaking research on new materials to improve battery performance, leading directly to the materials that we use today. The third awardee, Akira Yoshino from Japan, was responsible for demonstrating the commercial viability of lithium-ion batteries. Their research, which was conducted independently, spanned over 50 years from the fledgling studies of lithium-ion batteries in the 1970s to seeing that technology become a global $33 billion market in the year the Nobel Prize was awarded.

Before we look in more detail at each of the three Nobel Laureates and how their contributions got to the point that almost every single mobile device from electric cars to smartphones, from power tools to cordless vacuum cleaners is powered entirely by the movement of tiny lithium ions over a distance less than half the thickness of a human hair, let's consider the basic components of a lithium-ion battery and the history that preceded this most versatile of energy storage technologies.[2]

Every lithium-ion battery is comprised of four basic components—two electrodes (a cathode and an anode), an electrolyte, and a separator. Each of these components plays an essential role in enabling the battery to produce a reliable and safe electric current.

The cathode provides a source of lithium ions and determines both the capacity and voltage of the battery. Because lithium is unstable in its elemental form it is present in the cathode as a lithium-containing oxide mineral, which is referred to as the "active material." In simple terms, the greater the amount of lithium stored in the cathode the more work the battery can do. The larger the potential difference between the cathode and the anode the higher the overall voltage of the battery. A higher voltage also increases the capacity of the battery, extending battery life while delivering more power. So, the cathode determines many of the aspects of battery performance; how far you can drive an electric vehicle between plug ins or how long you can use a smartphone before you need to find a charging outlet.

A closer look at the cathode of a commercial lithium-ion battery shows that it is actually comprised of a mixture of materials. The active material is applied to a thin aluminum or copper foil together with a conductive additive to improve the electrical conductivity of the cathode assembly. A binder acts as an adhesive to hold everything together making sure the active material is firmly adhered to the metal foil. Carbon nanotubes, which were described in Chapter 5 are an example of an excellent electrical conductor that can be added to the cathode formulation. An example of a binder is styrene butadiene rubber (SBR), a form of synthetic rubber.[3]

Just like the cathode, the anode substrate is also coated with an active material, which captures lithium ions coming from the cathode, traps them, then releases them back to the cathode. When the battery is being charged, lithium ions are stored in the anode. When the battery is being used (the discharge state) lithium ions flow back to the cathode through the electrolyte. Electrons that have been separated from the lithium ions generate an electric current as they move through an external circuit that connects anode and cathode.

The anode material found most often in lithium-ion batteries is graphite, one of the many allotropic forms of carbon. Graphite is combined with a conductive additive and a binder before being coated onto a copper substrate. Graphite has the layered structure shown earlier in Fig. 5.3 with a separation between the individual graphene sheets of 0.34 nm. Tiny lithium ions, which are only 0.15 nm across, can comfortably fit between the sheets until they are required to move back to the cathode. Not only is graphite able to house the lithium ions with minimal distortion, it is stable in contact with the liquid electrolyte inside the battery environment. Graphite is also inexpensive, an important consideration when selecting materials for high volume products such as lithium-ion batteries.

Cathode and anode are separated by an electrolyte—a liquid that allows the passage of lithium ions but will not conduct electrons. Forcing the electrons to flow through the external circuit, rather than through the electrolyte, is what allows them to do work. How fast the lithium ions shuttle between the anode and cathode depends on the composition of the electrolyte. We want the diffusion rate to be as high as possible to enable rapid charging. The electrolyte itself is a complex mixture, composed of salts, solvents, and various additives. Of the salts the most widely used is lithium hexafluorophosphate ($LiPF_6$), a good conductor of lithium ions. Organic liquids that can readily dissolve the salts to form a uniform solution are used as solvents. Additives are selected to modify specific properties of the electrolyte mix. One important role additives play is as flame retardants to improve battery safety. Flammability has been a persistent problem associated with lithium-ion batteries since their inception. During the period March 1991–January 2018 the United States Federal Aviation Administration (FAA) reported over 200 instances of fires or explosions due to lithium-ion batteries in the air or in airports. In 2016, 92 Samsung Note 7 smartphones caught fire, resulting in a mass product recall.[4]

The function of the separator is simple, but nonetheless important. It is a physical barrier that keeps the anode and cathode apart. Together with the electrolyte, the separator plays an active role in determining the safety of the battery. Commercial separators are porous polymer films usually made of low-cost plastics including polyethylene and polypropylene.

The potential of lithium as an anode electrode in a battery was demonstrated in a 1913 publication by Gilbert Lewis who had at the time recently joined the University of California Berkeley and Frederick Keyes at the Massachusetts Institute of Technology.[5] To measure the electrochemical properties of lithium the two chemists constructed a cell that consisted of one electrode of metallic lithium, the other a lithium-mercury amalgam containing 0.035% of lithium. The electrolyte was propyl amine nearly saturated with lithium iodide. Propyl amine was chosen because it would not readily dissolve the lithium metal, but it was a good solvent for the lithium salt. Using this cell Lewis and Keyes measured the electrode potential for lithium and found that at 3.3044 V it was the highest value they had measured so far, higher than two of the other alkali metals sodium at 2.9981 V and potassium at 3.2084 V. The high electrode potential for lithium is one of the reasons that it is such a valuable battery electrode.[6]

Primary batteries (single use only, non-rechargeable) featuring a lithium metal anode partnered with several different cathodes became commercially available beginning in the late 1960s, initially for space applications then for more widespread consumer use.[7] In 1971, Japanese company Matsushita Electric used a polycarbon monofluoride, also called graphite fluoride $(CF_x)_n$, cathode for their first lithium primary battery, whilst Sanyo developed a battery with a ceramic cathode of manganese dioxide, MnO_2. These batteries found their way into a wide range of electronic equipment including calculators, watches, and cameras. Since 1972, a battery comprising a lithium metal anode, a lithium iodide electrolyte, and a polyvinylpyridine cathode has been used in heart pacemakers because of its excellent reliability, predictable discharge rates, and safety. All the requirements we would want in a device controlling our heartbeat. Millions of these life-saving devices have been implanted in patients throughout the world.[8]

The reason these batteries are nonrechargable is that over time the formation of lithium iodide (LiI) by the reaction between the lithium anode and iodine causes the resistance of the cell to increase and the battery begins to run down eventually needing replacement. Pacemaker batteries can last as long as fifteen years. In the case of manganese dioxide cathodes, when the lithium moves from the anode it reacts with the manganese dioxide forming the compound lithium manganese dioxide, $LiMnO_2$. This reaction is not readily reversible because manganese, which is now in a 3+ oxidation state rather than the original 4+ state destabilizes the lattice structure of the cathode.[9] As a result, once the reaction is complete the battery needs to be replaced.

To make a secondary—rechargeable—lithium battery it is necessary that the lithium ions can go into and out of the electrode structure many thousands of times with destroying it. These types of reversible reactions are called intercalation reactions.[10] Intercalation compounds have channels that can accommodate the tiny lithium ions. One of the first intercalation compounds considered for secondary lithium batteries was proposed by Brian Steele a materials scientist at Imperial College, London at a NATO conference held in Italy.[11] Steele suggested the use of transition metal disulfides as intercalation electrodes. Transition metal disulfides including tungsten disulfide WS_2, titanium disulfide TiS_2, and molybdenum disulfide MoS_2 have layered structures similar to that of graphite. Rather than layers of carbon atoms held together by weak van der Waals bonds, the disulfides are composed of blocks of sulfur-metal-sulfur or S-M-S sheets stacked on top of each other. It is these sheets that are held together by van der Waals bonds. The spacing between sheets is on the order of 0.62 nm, which can accommodate the much smaller lithium ions (0.15 nm).[12]

The idea of using intercalation compounds such as transition metal disulfides opened up the possibility of a rechargeable lithium battery.

In 1976 M. Stanley Whittingham a scientist who had arrived at the Corporate Research Laboratories of Exxon in Linden New Jersey from California's Stanford University four years earlier patented a new battery system combining lithium metal as the anode active material and a titanium disulfide cathode.[13] The reaction to form the lithium titanium disulfide intercalation compound was found to be very fast and highly reversible without destroying the layered structure of the host material. Whittingham showed that his lithium battery could complete over 1000 charge and discharge cycles with very little degradation in performance.

Unfortunately, what makes lithium such an excellent active material for a battery, its reactivity, its enormous drive to release its lone outer electron, also creates drawbacks in terms of stability.[14] While Whittingham's design was functional—the battery had a high voltage and produced an electric current—it was soon found to have some technical problems. Early attempts at battery production suffered considerable setbacks, even having to call out the fire brigade to deal with fires caused by exploding cells.

The issue was tiny lithium whiskers, called dendrites, that would grow out from the lithium anode. When the dendrites reached the titanium disulfide cathode there would be a short-circuit, which could lead to an explosion. Modifications to the lithium anode by the addition of aluminum and changing the electrolyte chemistry improved battery safety and small-scale production of the battery began in 1978 at the Exxon Enterprises Battery Division. In a recent study, some of these first commercially available rechargeable lithium batteries were tested after being in storage for 35 years.[15] They still worked! Retaining more than 50% of their original capacity.

Unfortunately, global circumstances would interfere with further development of Whittingham's rechargeable lithium battery. The early successes were not sufficient to sustain battery development at Exxon. With the dramatic drop in the price of oil beginning in the early 1980s the company was forced to make cutbacks and one of the programs that was discontinued was battery development. Whittingham's patent was licensed to other companies who took up the mantle of hoping to achieve the ultimate goal of a lithium battery powering an electric car.

One person who was aware of Whittingham's revolutionary battery and had an interest in the internal structure of oxides was Oxford University chemist John Goodenough. Goodenough was born in Jena Germany in 1922 to American parents. He studied mathematics at Yale University before serving during World War II as a meteorologist in the US Army. After the war, Goodenough moved to the University of Chicago where he was awarded his doctorate in physics in 1952. After more than two decades at MIT's Lincoln Laboratory Goodenough moved to Oxford University as head of the Inorganic Chemistry Laboratory where he began a systematic search for a metal oxide that could replace Whittingham's transition metal disulfide as a cathode.

Goodenough expected that a metal oxide would produce a higher voltage than the sulfide when intercalated with lithium ions, but that alone was not enough to make a good cathode. The structure of the oxide would need to be stable when the ions were reversibly, and repeatedly, moved in and out. In 1980 the critical discovery came when Goodenough together with Koichi Mizushima, Philip Jones, and Philip Wiseman identified a cathode material that enabled development of the rechargeable lithium-ion battery in a form very similar to that produced by the billions today. This breakthrough ushered in the age of portable electronic devices.[16]

The cathode that came from the Oxford group was lithium cobalt oxide, $LiCoO_2$, which produced a voltage almost twice that of the lithium disulfide it replaced.[17] Lithium cobalt oxide has a layered structure, which is illustrated

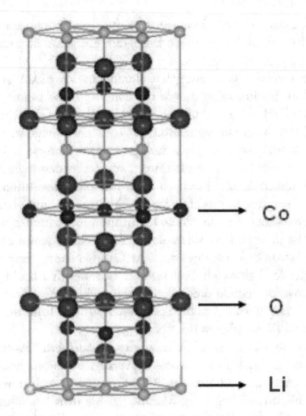

Fig. 6.1 The crystal structure of the layered form of lithium cobalt oxide. The oxygens (red) form what is called a close-packed arrangement stacked in a repeating ABCABC sequence. Cobalt (blue) and lithium (green) ions occupy alternating layers in the structure (Reprinted from Xia H, Meng YS, Lu L, Ceder G (2007) Electrochemical behavior and Li diffusion study of $LiCoO_2$ thin film electrodes prepared by PLD. Article available at https://dspace.mit.edu/bitstream/handle/1721.1/35827/AMMNS001.pdf?sequence=1&isAllowed=y. Accessed 11 August 2022)

in Fig. 6.1. The oxygen ions are stacked in a close packed arrangement that repeats every three layers giving an ABCABC-type stacking sequence. Cobalt and lithium ions occupy alternating layers between the oxygens. During charging and discharging the lithium ions hop from one site in the crystal structure to an adjacent one.

The structure of lithium cobalt oxide also suggested other compositions that might work as cathodes for lithium-ion batteries obtained simply by substituting cobalt with other transition metals such as nickel or manganese. For instance, the compound lithium manganese oxide, $LiMn_2O_4$, was considered for several reasons: manganese is considerably less expensive than

cobalt; manganese does not have some of the same geopolitical issues associated with its acquisition (more on this later); and the manganese variant is
more stable at high temperatures.

Another avenue that Goodenough explored, now a professor at the University of Texas at Austin having moved there in 1986, was lithium-ion battery
cathodes based on the structure of the mineral olivine, in particular the
compound lithium iron phosphate, $LiFePO_4$.[18] The olivine structure is very
common in nature, with perhaps the best-known example being the light
green gemstone peridot. There were now two parallel development paths for
new cathode materials, one based on additions and substitutions to lithium
cobalt oxide the other on modifications to lithium iron phosphate.

While developments in materials for lithium-ion batteries were coming
from the West there was considerable interest in deploying this technology
in Japan. Electronics companies such as Casio, Nikon, Sanyo, and Sony
were desperate for lightweight rechargeable batteries that could power their
range of innovative mobile products including video cameras, cordless tele-
phones, and computers. It was here that Akira Yoshino from the Asahi Kasei
Corporation in Tokyo adds to the story.

Akira Yoshino was born in Suita, a city in northern Osaka Prefecture.
After studying technology at Kyoto University, he began working at the
Asahi Kasei chemical company in 1972. Yoshino stayed at the company
throughout his career eventually heading up his own laboratory. In 1985
Yoshino patented a lithium-ion battery comprising Goodenough's cathode
paired with an anode of petroleum coke, a carbon material that at the
molecular level has spaces that can house lithium ions.[19] This was the first
demonstration of a commercially viable lithium-ion battery. It was stable,
lightweight, had a high capacity, and produced 4 V (a very high value for this
type of cell.)

Because the lithium ions can rock or shuttle back and forth between the
carbon anode and the oxide cathode and vice versa without changing the
integrity of either electrode the battery was truly rechargeable demonstrating
a lifetime of many hundreds of discharge and charging cycles. A critically
important feature of Yoshino's battery was that it was much safer than the
earlier batteries made using a pure lithium metal cathode. The significance of
this advantage was stated by Yoshino on successful completion of the safety
tests, "[this was] the moment when the lithium-ion battery was born."[20] In
1991, a successful commercial lithium-ion secondary battery was developed
by a joint Sony and Asahi Kasei team led by Yoshiro Nishi.

Development and eventual commercialization of the lithium-ion battery
pioneered by Stanley Whittingham, John Goodenough, and Akira Yoshino

created a revolution in portable electronics including an array of smartphones, tablets, laptop computers and wearable devices that has without doubt changed the world. Lithium-ion batteries are also staking a strong claim to be the power source of choice for the rapidly growing industry of electric transportation, from electric cars, buses, scooters, and trucks to more electric airplanes like Boeing's 787 "Dreamliner." Each battery pack on the Dreamliner contains eight 2.5–4.025 V cells providing a total voltage in the range 20–32.2 V. With a pack capacity of 75 amp-hours the batteries help start the auxiliary power unit when the plane is on the ground and provide a backup for electronic flight systems. In comparison, the size of a cell that powers small portable electronic devices may be in the range of just 1 to 4 amp-hours.[21]

In 2019, the year that the Nobel Prize in Chemistry was awarded in recognition of the development of the lithium-ion battery, annual sales of electric cars exceeded 2.1 million, a record-setting year boosting the total worldwide to 7.2 million vehicles. Almost 50% of these electric vehicles were in China. And with projected year-on-year growth of 40% electric cars are going to be an increasing part of the global car stock.[22] A major contribution to this rate of growth is new mass-market models such as the Tesla Model 3, with over 440,000 vehicles delivered in 2020 alone. Supercar makers including Lamborghini are moving away from polluting and inefficient internal combustion engines. The Italian company announced plans to produce only hybrid electric vehicles by 2024 and have all-electric vehicles ready by the end of the decade. Market rival Ferrari has promised to launch an all-electric model by 2025—five years ahead of Lamborghini and on a similar timeline to the luxury Jaguar brand.

A recent study published by the International Monetary Fund considered the impact of different adoption rates of electric vehicles over the next two decades. In the fast-adoption model over 90% of all vehicles on the road could be electric by 2042.[23] All these vehicles will require batteries; lots and lots of them.

For most electric vehicles including the Tesla Model 3, the battery of choice is the lithium-ion battery. Its high energy density and low weight are two of the properties that distinguish lithium-ion batteries from other battery chemistries. In 2020 total annual demand for lithium-ion batteries was about 300 gigawatt-hours. By 2030 that number is projected to have grown to more than 2,000 gigawatt-hours with 1,500 gigawatt-hours just for passenger electric vehicles.

The projected size of the global electric vehicle market, which already exceeds that for consumer electronics is the driving force behind expansion of battery manufacturing capacity—both in the United States and worldwide—such as the sprawling Tesla Gigafactory in Sparks, Nevada, which has a planned annual production capacity of 35 gigawatt-hours. That is enough for about 500,000 electric cars. In turn, growth in battery manufacturing has led to cheaper and cheaper batteries. In 2010 a lithium-ion battery pack cost $1,160 per kilowatt-hour. Ten years later that number was less than $200 per kilowatt-hour. A decrease of 80% over the past decade!

Despite the amazing success of the lithium-ion battery there remain significant challenges. For instance, a smartphone battery needs to be charged daily. More frequent charging is required if movies are being streamed, satellite navigation is turned on, or multiple applications are running simultaneously. The hunt for an unused charging station at an airport is a common occurrence to get that final boost of lithium ions into the anode before leaving for a long-haul flight.

Even a state-of-the-art electric vehicle cannot drive from the rolling hills of the Palouse to the waters of the Puget Sound without the driver needing to find a charging station somewhere around Ellensburg or North Bend. By 2030, battery electric vehicles are assumed to reach an average driving range of 350–400 km, still at least 60 km short of my trip across Washington state. The automobile industry continues to look for the perfect battery technology capable of providing the long-term energy storage necessary for electric vehicles to compete with the convenience of conventional gasoline cars. At the same time, manufacturers such as Panasonic Sanyo, CATL, and LG Chem continue making incremental improvements to eke out every possible mile or hour of battery life. So that prompts two questions:

- In the short term: what is the next step in lithium-ion battery technology?
- In the longer term: is lithium-ion chemistry the future for powering mobile applications?

To answer the first question researchers around the world are searching the Periodic Table of Elements to look for modifications to the composition of layered oxides including lithium cobalt oxide $LiCoO_2$, oxides with the spinel structure like lithium manganese oxide $LiMn_2O_4$, or compounds in the lithium iron phosphate, $LiFePO_4$, family that have the olivine structure.[24] The goal is to produce a cathode that enhances battery performance, further lowers the cost and reduces the reliance on critical raw materials. That is, materials that have been identified by the United States Government and other organizations including the European Union as serving essential economic, technological, or national security functions and where the supply chain is vulnerable to disruption.[25]

The lithium cobalt oxide material developed by John Goodenough and his group remains one of the best cathodes to date because of its high operating voltage. But the high cost and limited capacity of lithium cobalt oxide have driven efforts to replace cobalt with other transition metals, in particular ones that are more abundant and environmentally benign. A combination that has proved commercially successful is nickel manganese cobalt oxide (known by the abbreviation NMC), which has become the most common cathode chemistry used in electric vehicles. It is also increasingly the battery of choice for power tools and electric battery-assist bicycles (already a 3.7 million units per year market and growing rapidly.) A particular advantage of the NMC composition is that the nickel content can be increased while the cobalt content is lowered or, in some cases, even eliminated entirely.[26]

Much of the performance and certainly a large part of the cost of a lithium-ion battery is determined by the cathode material. On the anode side of the cell there is also a great deal of research effort to look for materials that have higher lithium capacities than the current anode of choice, graphite. The options appear more limited than they are for the cathode because of the need to identify a material that can accommodate a large reversible volume change when the lithium ions enter the structure during charging (lithiation) and then leave during discharging (delithiation). When lithium ions intercalate between the graphene sheets of the graphite structure the volume change is only 10%, which can quite easily be accommodated. But graphite anodes have a poor energy density, which means that discharge time is limited. For instance, we know that it is necessary to charge our smartphones every day, rather than a preferred once a week or even once a month.

Two materials that have shown some promise towards improving the performance on the anode side of a lithium-ion battery, but have so far had limited commercial impact, are tin and silicon.

The theoretical energy density of tin is over 2½ times greater than that of graphite but comes at the cost of an enormous 252% volume change in the material between charged and discharged states. Early attempts at making lithium-ion batteries using tin anodes proved unsuccessful largely because the anode, which was in the form of a thin film would crumble gradually losing its structural integrity until it eventually was able to hold no charge at all.[27] Suggestions of ways to reduce the impact of the large volume change in the anode included reducing the particle size of the anode material until it was in the nanoscale regime. Research has shown that if the anode material is formed into a nanowire that large volume changes can be more easily accommodated than when the material is in the form of a film or a powder.

Figure 6.2 shows an example of tin nanoneedles electroplated onto a copper substrate before being inserted into a battery.[28] Nanoneedles vary between 1 and 5 μm in length and are tapered with the tips being 20–100 nm across and the base width between 50 and 300 nm. Once integrated into a battery tin nanoneedles were shown to go through 100 lithiation/delithiation cycles while still retaining their needle-like shape. While the tin anodes cycle, they have not been able to reach energy densities close to the theoretical value. A further concern with the use of tin, which does not impact other possible anode materials for lithium-ion batteries is that tin is one of the so-called 3TG minerals. These are the "conflict minerals"—tin, tantalum, tungsten, and gold—that originate from the Democratic Republic of Congo or adjoining states.[29] Acquisition of the minerals that are processed into the 3TG metals is associated with financing armed conflicts in the region and human rights violations linked to child labor and death. Determination of the specific source of these raw materials is critical to the development of a complete and transparent conflict-free mineral supply chain.

The primary source of tin is the oxide mineral cassiterite, SnO_2. In descending order China, Indonesia, Myanmar, Peru, Bolivia, and Brazil, were the leading producers of tin in 2020.[30] China and Indonesia together account for more than half of the world's supply. Although the Democratic Republic of Congo and surrounding countries accounted for less than 10% of the global supply of tin in 2020, estimated deposits in this central African country are substantial. Mineral reserves of tin in the United States are small. With neither domestic mining nor smelting of tin alloys happening in more than twenty years, there is complete reliance on foreign imports of tin, which although not widely used in lithium-ion batteries is important for many applications in the electronics industry including as a component in solders and in the transparent electrical conductor indium tin oxide used in, for instance, flat panel displays.

An active anode material that has an even higher capacity than tin and almost ten times that of graphite is silicon. Silicon offers an extremely high energy density, which unfortunately comes at the cost of an enormous 320% volume change during cycling. This is more than enough to damage the brittle anode structure from repeated expansion and contraction as the lithium ions are inserted then extracted. Research from a group at Stanford University in California showed that using silicon in the form of nanowires helped decrease degradation of the anode allowing it to accommodate the large strains during lithium cycling.[31] The silicon nanowires were grown by a

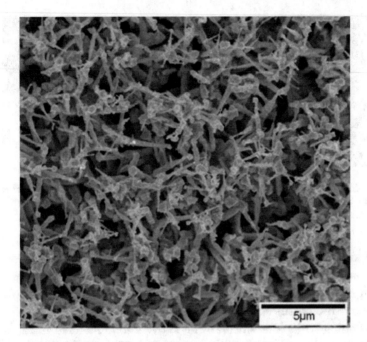

Fig. 6.2 Scanning electron microscope image of tin nanoneedles electrodeposited onto copper to form an anode for a lithium-ion battery (Reprinted from Mackay DT, Janish MT, Sahaym U, Kotula PG, Jungjohann KL, Carter CB, Norton MG (2014) Template-free electrochemical synthesis of tin nanostructures. Journal of Materials Science 49:1476–1483 with permission from Springer Nature)

vapor-liquid-solid (VLS) process that requires a controlled environment with the absence of all oxygen to avoid high temperature oxidation of silicon.[32] This constraint together with cost, limit the ability to scale up the process to commercially viable levels even though the performance of the battery was promising.

Using a different processing method Stanford University spin-out company Amprius has partnered with European airplane manufacturer Airbus to provide lithium-ion batteries made using silicon nanowire anodes combining with solar energy to power the Airbus Zephyr S unmanned aircraft.[33] The patented silicon nanowire structures shown in Fig. 6.3 are grown directly onto the metal substrate without the need for any binder or additive. The process is very similar, but on a much smaller scale, to the growth of polysilicon rods used to make the high-purity material that is eventually turned into large single crystals for the semiconductor industry.

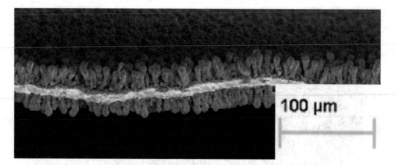

Fig. 6.3 Scanning electron microscope image showing the cross section of a silicon nanowire anode. Note that the silicon nanostructures are formed on both sides of the metal substrate (Image courtesy of Amprius)

As a demonstration of battery performance, using the silicon nanowire anode, on its maiden flight the Zephyr S flew for 26 days at an average altitude on the edge of the ozone layer in the stratosphere (21 km; about 13 miles above the Earth's surface).

Continuing to look at silicon as an active anode material makes a lot of sense from a materials availability perspective. Silicon is the most common element on Earth after oxygen, accounting for almost 30% of the minerals that make up the crust.

An abundant source of silicon is diatomaceous earth—diatomite—a sedimentary rock that has been deposited over millions of years. It is composed of the deceased, fossilized frustules—cell walls—of diatoms. The frustules are principally silica, SiO_2, and exist in strikingly beautiful and many varied forms as illustrated by just one example in Fig. 6.4. There are massive deposits of diatomaceous earth that spread over thousands of square miles and are thousands of feet deep. The United States is the world's largest producer of diatomite, well ahead of Denmark, Turkey, and China.[34] Domestic production in 2020 was estimated to be 770,000 tons with a processed value of around $260 million. Diatomaceous earth is an abundant, inexpensive, and easily accessible source of silica. Recently it has been demonstrated that diatoms make ideal templates to convert into highly porous nanoscale silicon, which can be incorporated directly into lithium-ion battery anodes.[35] In one method the silica is reduced in the presence of magnesium to nanosilicon, which is then coated with carbon to form the active anode material. Although the measured battery capacities were below the theoretical value and much more testing would need to be done before diatomaceous earth-derived material would make it into commercial batteries the approach demonstrates an

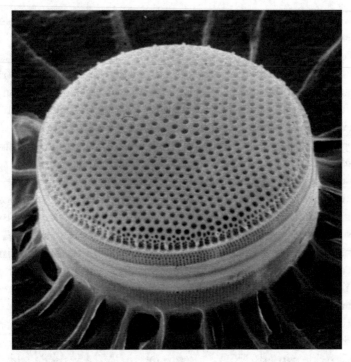

Fig. 6.4 Scanning electron microscope image of a centric—radially symmetric—diatom. Centric diatoms have a wide range of shapes and sizes typically between 20 and 200 μm in diameter (Image courtesy of CSIRO)

example of developing materials using inspiration or templates derived from nature.

With graphite having reached its energy limit and metallic alternatives such as tin not able to compete in terms of theoretical performance, continuing to look at how we might use silicon as an anode material in a lithium-ion battery certainly warrants further consideration and more research.

In the hunt for increasingly better batteries, no one has yet succeeded in inventing something that beats the lithium-ion battery's high capacity and voltage. Certainly, for the next decade, the lithium-ion battery is set to dominate the electric vehicle market. But what about beyond 2030 and further into the twenty-first century when almost every automobile manufacturer will have a full range of electric vehicles? Are there alternate chemistries that might make an impact? Stanley Whittingham, one of the pioneers of the lithium-ion battery, recently reviewed many of the different beyond-lithium alternatives.[36] Whether continuing down Group I of the Periodic Table with sodium and potassium or moving across to Group II with magnesium, and calcium, there are significant limitations and challenges. Whittingham suggests that the most likely replacement for lithium-ion batteries are sodium-ion cells, which have also been around since the early 1970s.

Structurally a sodium-ion battery is closely related to its lithium counterpart. Even some of the materials are similar. But the chemistry is very different because it involves the movement of sodium ions. In a sodium-ion battery the anode is an amorphous form of carbon known as "hard carbon". Being larger than lithium, sodium ions are not able to comfortably fit between the ordered graphene layers in the graphite structure. It is for this reason that the disordered form of carbon is used where there are large enough spaces to incorporate the sodium ions.

The cathode in a sodium-ion battery is often a layered transition metal oxide for instance sodium cobalt oxide, $NaCoO_2$. Separating the two electrodes is a liquid electrolyte and a porous polymer separator. A typical electrolyte, which must be a good conductor of sodium ions consists of an organic carbonate containing a sodium salt such as sodium fluorophosphate, $NaPF_6$. You may recognize this as being very similar to the electrolyte mixture in a lithium-ion battery.

A head-to-head comparison shows that the specific energy of current state-of-the-art lithium-ion batteries is 285 Wh/kg using a lithium nickel cobalt aluminum oxide cathode NCA.[37] Replacing the NCA cathode with lithium iron phosphate, $LiFePO_4$, reduces the specific energy density to around 130 Wh/kg. This value is in the range predicted for fully developed sodium-ion batteries.[38] So, the current conclusion is that sodium-ion batteries would be suitable for applications similar to those where lithium-ion batteries with lithium iron phosphate cathodes are used. These applications may include storage for intermittent renewable energy sources including solar and wind, power back-up units for electric utilities, and home storage use. But the significantly lower energy densities would not be acceptable for most mobile applications, except for short-range electric vehicles where there is currently a market in China.

At the moment, and into the foreseeable future, lithium-ion batteries dominate mobile applications creating a demand for the requisite raw materials, which include not only lithium, but cobalt, nickel, manganese and other metallic elements that are used to fabricate the cathode. The estimated material demand for the lithium-ion batteries used in the electric vehicles sold in 2019 was about 19,000 tons of cobalt, 17,000 tons of lithium, 22,000 tons of manganese and 65,000 tons of nickel. By 2030 the annual demand for these ingredients is projected to increase dramatically: 180,000 tons of cobalt, 185,000 tons of lithium, 177,000 tons of manganese, and 925,000 tons of nickel. Depending on vehicle sales the actual numbers could turn out to be even higher.

Lithium is the first and lightest metal in the Periodic Table of Elements. In the early 1970s lithium was a "relatively obscure mineral commodity" that was of only limited commercial interest.[39] The major applications included lithium greases as lubricants in aircraft engines and lithium-based soaps. During the Cold War, the lightest lithium isotope lithium-6 (^{6}Li) was used for producing tritium (^{3}H) and the compound lithium deuteride, which was used as a fusion fuel in nuclear weapons. At the time the United States led the world in lithium production. Since then, the global picture for lithium has changed considerably. It is now one of the most coveted raw materials in the world and American production lags far behind that of the major producers Australia, Chile, and China.

Lithium has been listed as one of the critical or near-critical elements largely based on its importance in green technologies, specifically lithium-ion batteries and in lightweight aluminum–lithium alloys for aerospace that can lead to significant savings in fuel consumption when compared to other alloys.

Lithium is present in the Earth's crust at an abundance of 0.0020%.[40] It is more common than lead (0.0014%), tin (0.0002%), and tungsten (0.0001%), but extraction is difficult because it is so thinly distributed. As with most metals, lithium is not found in the metallic, or native, form in nature. Rather it is bound with oxygen in complex silicate minerals such as petalite ($LiAlSi_4O_{10}$), the complex pink mica mineral lepidolite ($K(Li,Al)_3(Al,Si,Rb)_4O_{10}(F,OH)_2$), and spodumene ($LiAl(SiO_3)_2$). In addition to the mineral sources of lithium, over half of the world's lithium is in brine deposits, salty groundwater enriched in dissolved lithium chloride. The United States Geological Survey estimates global lithium reserves at about 86 million tons.

Global production of lithium in 2020 was a little over 80,000 tons, with Australia, Chile, and China accounting for almost 90% of that amount.[41] Chile was the leading producer of lithium from brine, while Australia led the way in extracting the metal from pegmatites, hard rock mineral deposits. Located 160 miles south of Perth, in the south-west of Western Australia, the Greenbushes Mine is the world's largest operating lithium mine with an annual production capacity equivalent to 30,000 tons of lithium metal. Currently more than 56,000 tons of lithium are used annually in the manufacture of lithium-ion batteries. By 2027 this number is expected to reach well over 200,000 tons.[42]

With anticipated demand for electric vehicles to continue to increase well into the twenty-first century there will be even further demands on lithium supplies. One way to help in sustainably meeting these demands is the development of lithium recycling programs. By 2030, it is estimated that close to one million lithium-ion batteries will have reached their end of life.[43] At this time less than 5% of lithium-ion batteries are recycled to recoup the lithium and other valuable metals including cobalt and nickel. This is in stark contrast, for example, to lead-acid batteries that are used in all non-electric cars to operate the starter motor and to power the vehicle's electrical circuits where virtually all of the metal is recycled.

The National Renewable Energy Laboratory (NREL), a government funded research facility in Golden Colorado, identified twenty-three companies that recycled lithium-ion batteries from electric vehicles in a paper published in 2019.[44] Twelve of those companies were in Europe, five in North America, and six in Asia.

Retriev Technologies has recycled lithium metal and lithium-ion batteries since 1992 at its facility in Trail, British Columbia just north of the United States-Canada border. In 2015, the company began operating the first domestic recycling facility for lithium-ion vehicle batteries in Lancaster, Ohio. The total recycling capacity of both plants is 8,500 metric tons per year. For many of the companies offering recycling services for lithium-ion batteries their interest is not primarily in the lithium itself, but rather, as mentioned earlier, in the nickel and cobalt contained in the cathodes. In the case of Retriev, the recycled lithium, in the form of a brine, is generally sold to a steel manufacturer rather than going back into batteries. This approach mimics the situation encountered in much of plastic recycling where the material is downcycled into a lower grade product. For instance, a recycled polyethylene terephthalate water bottle ends up not in a brand-new bottle but in a park bench or as fleece fibers for clothing.

Recycling spent lithium-ion batteries involves multiple steps. First the battery must be disassembled. Because components in the battery can catch fire special handling precautions are required to reduce the risk of combustion. Once the batteries have been shredded the next step is to extract the valuable components from the "black mass." There are two methods that are presently used in industry: pyrometallurgy and hydrometallurgy.

Pyrometallurgy involves heating the black mass to a temperature sufficient to melt the metals. The liquid can then be removed and allowed to cool leaving a solid mixture containing all the metallic components of the battery. From this mixture the individual metals—particularly cobalt—can be separated. Separation of the individual elements is not a trivial process requiring a lot of energy. During heating, the graphite anode material is destroyed while the lithium ends up in the waste slag, which frequently means that it is not saved for recycling. Any plastic components in the black mass act as fuels during heating, which helps reduce the overall energy input necessary to maintain the high temperatures inside the reactor. The exhaust emissions are toxic and require scrubbing before being released into the atmosphere. For one ton of batteries, 5000 megajoules of heat are needed.[45] For a comparison, this is equivalent to the energy required to brew 17,000 pots of coffee.

An alternative approach to recycling the components in a lithium-ion battery is hydrometallurgy. Hydrometallurgy uses strong acids or other powerful solvents to leach out the metals from the battery including the lithium, which is often not recovered in pyrometallurgy. A further benefit to hydrometallurgy is that because the process is conducted at room temperature energy costs are lower and it is possible to recover the graphite from the anode. The primary drawback with hydrometallurgy is it generates a toxic and corrosive waste-water stream that needs to be treated, which then adds to the overall expense of the process.

Neither pyrometallurgy nor hydrometallurgy currently recycles the electrolyte, which is destroyed during processing. Balancing the pros and cons, hydrometallurgy appears to be, at present, the preferred technique to reclaim the active materials from lithium-ion battery scrap. But as identified by Ahmad Mayyas and colleagues at NREL recycling lithium-ion batteries faces some long-term challenges. For instance, there may very likely be changes in the composition of the active materials, a different mix of transition metals, which could render current recycling processes "obsolete or ineffective." If cobalt concentrations in the cathode continue to decrease or are even eliminated completely then the economic viability of lithium-ion battery recycling may be threatened.

The ability of hydrometallurgy to recover more of the active materials, including the lithium, further supports the claim that it will be the preferred recycling method even if battery chemistries continue to evolve. As with all current recycling operations whether glass, plastic, metal, paper, they only really work when the cost of the "virgin" raw materials is equal to or greater than the economically viable price for recycled materials.

As mentioned several times in this chapter in addition to lithium another critical element in current state-of-the-art lithium-ion batteries is cobalt. Cobalt has been an important component since John Goodenough's discovery of the properties of lithium cobalt oxide as an active cathode material. Most electric vehicles on the road today rely on cobalt. And as demand for electric vehicles increases there will be an even greater demand for this metal.

Great Britain's Royal Society of Chemistry rates cobalt as one of the critical materials in lithium-ion batteries giving it a supply risk of 7.6 out of 10. A high risk; equivalent to that of the precious metal platinum.[46] One of the challenges of working with cobalt is that 70% of the world's mine production is supplied by companies and organizations in the Democratic Republic of Congo. As widely report, the country is rife with corruption. It has experienced decades of an unstable political environment and civil wars leading to cost fluctuations in mined materials and supply chain disruption. The country also has a dreadful human rights record with mining operations linked to child labor, extortion, abuse, and dangerous working conditions that can often turn deadly.

Almost 100,000 tons of cobalt were produced in the Democratic Republic of Congo in 2020 and the country has over half the world's reserves.[47] A recent analysis led by Elsa Olivetti at the Massachusetts Institute of Technology has found that demand for cobalt could be as high as 430,000 tons by 2030. This represents more than four times world production just a decade earlier.[48] Major technology companies including Microsoft and Apple have acted upon the supply risk of cobalt with growing demands in the last few years to work with suppliers and miners to increase the sustainability of the supply chain and ensure respect for human rights.

To mitigate potential shortages of cobalt more efficient and widespread recycling programs are necessary. Alternatively, the demand for cobalt can be reduced or even eliminated by replacing part or all of it in cobalt-free lithium-ion batteries. Lithium iron phosphate is a cobalt-free active cathode material that is being used for instance in short-range electric vehicles aimed primarily at the Chinese market. Lithium iron phosphate batteries are less expensive and have a lower environmental impact compared with those requiring cobalt.

While predicting the future of technology is often fraught with uncertainly it is difficult to imagine that lithium will not feature highly in the short, medium, and long terms as a primary constituent in rechargeable batteries for our range of mobile devices and electric vehicles. Lithium boasts the lowest reduction potential of any element, giving it the highest possible cell voltage. It is the lightest of all the metallic elements and with an ionic radius of just 0.076 nm is the smallest of all the singly charged ions. These properties alone make lithium nearly ideal as an active material in lightweight, energy dense batteries ensuring its future role.

Cobalt and other transition metals including nickel will continue to be used in lithium-ion batteries as active cathode materials because they form oxide crystal structures that can accommodate large fractions of lithium. If existing sources of cobalt, whether from mining or recycling, cannot satisfy demand then exploration of non-traditional reserves may accelerate. The United States Commercial Space Launch Competitiveness Act of 2015 makes it legal for companies to own and sell resources they extract from space, including the Moon and asteroids. Cobalt and other transition metals have been found in lunar rocks and soil samples brought to Earth from the Apollo missions in concentrations very similar to those in the Earth's crust. Cobalt and nickel are often associated with iron-rich minerals known as siderophiles. If bringing substantial amounts of materials back from the Moon or capturing and mining asteroids proves too difficult and expensive then rather than looking towards space we may look to the ocean depths where there are vast deposits of critical minerals, precious metals, and rare-earth elements.

Interest in deep sea mining is not new. For instance, conservative estimates of undersea deposits of diamonds off the west coasts of South Africa and Namibia are at least 1.5 billion carats, of which as much as 95% are gem quality.[49] Although the potential richness of these deposits have been known since the early 1900s they have recently begun to be significantly mined. In 2018 the De Beers company extracted 1.4 million carats from the coastal waters of Namibia.

Cobalt occurs in metal-rich crusts that cover the slopes and summits of thousands of seamounts that rise a kilometer or more above the seafloor. Mining these rocky features is not only technically challenging but comes with massive environmental implications. The seamount environment is dominated by corals, anemones, feather stars, sponges, as well as tuna, sharks, dolphins, and sea turtles. Science journalist Olive Heffernan summarizes the potential detrimental environmental impact of deep-sea mining: "For most of the animals in the direct vicinity, mining will be lethal."[50]

If sourcing issues for cobalt are unresolved, then battery chemistries that either reduce or exclude entirely the use of this metal will be necessary. This is not as easy as it sounds as we are finding out with other attempts to substitute critical minerals with earth-abundant and more sustainable alternatives.

Despite the many challenges the future of mobility appears to be electric. One hundred years from now electric vehicles will be ubiquitous as governments throughout the world seek to meet their currently ambitious 100% zero-emission vehicle targets with the phase-out of internal combustion engine vehicles through 2050. In December 2019 France was the first country to put this intention into law with a target of a ban on the sale of all new gasoline cars by 2040. By that time at least seventeen other countries including Norway, South Korea, Belgium, and Austria will have banned gas-powered cars and by the middle of the century they will have been joined by over another twenty nations and many American states. A major challenge will be ensuring that we have sufficient and sustainable sources of raw materials to enable this emerging global electrification.

Notes

1. Goodenough JB (2009) Facts. NobelPrize.org. Nobel Media AB 2021. Mon. 3 May 2021. https://www.nobelprize.org/prizes/chemistry/2019/goodenough/facts/.
2. A typical distance between cathode and anode in a lithium-ion battery is 20–30 μm (mm).
3. SBR was developed as a replacement to natural rubber prior to World War II in Germany by chemist Walter Bock in 1929.
4. Kong L, Li C, Jian J, Pecht MG (2018) Li-ion battery fire hazards and safety strategies. Energies 11:2191.
5. Lewis GN, Keyes FG (1913) The potential of the lithium electrode. Journal of the American Chemical Society 35:340–344. In 1926 Lewis published a letter in the journal Nature coining the term "photon" for a quantum unit of light.
6. Lithium has the highest electrode potential of any element and is the third lightest making it nearly ideal for battery applications.
7. Primary batteries are "single use" and cannot be recharged. Secondary batteries are ones that can be recharged. It is secondary batteries such as lithium-ion batteries that are the most important for mobile applications.
8. Holmes C (2007) The lithium/iodine-polyvinylpyridine pacemaker battery—35 years of successful clinical use. ECS Transactions 6:1.
9. Zhu X, Meng F, Zhang Q, Xue L et al (2020) $LiMnO_2$ cathode stabilized by interfacial orbital ordering for sustainable lithium-ion batteries. Nature Sustainability. The oxidation state of manganese in MnO_2 is 4+ (each oxygen ion is 2−). In the compound lithium manganese dioxide, the manganese is now in a 3+ oxidation state.
10. The earliest use of the word "intercalation" as noted by M. Stanley Whittingham in his Nobel Prize acceptance speech is when the months of January and February were "intercalated" into earlier versions of the Roman calendar. It was only with the work of Whittingham and others that the word became applied to specific types of processes involving materials.

11. Brian Charles Hilton Steele was my doctoral advisor. His research interests included battery materials, solid oxide fuel cells, and high-temperature superconductors.

12. The interlayer spacing in MoS_2 is ~0.62 nm. If the sizes of the molybdenum and sulfur atoms are taken into account then gallery height is ~0.30 nm, which is a tight fit for lithium intercalation. Dong H, Xu Y, Wu Y, Zhou M et al (2018) MoS_2 nanosheets with expanded interlayer spacing for enhanced sodium storage. Inorganic Chemistry Frontiers 5:3099–3105.

13. Whittingham MS United States Patent 4,009,052. The patent application was filed on April 5, 1976 and issued February 22, 1977.

14. The oxidation potential is a measure of the reactivity of a metal. Lithium has an oxidation potential of +3.0401 V, the highest of any metal. As a comparison, gold which is considered an inert metal has a value of −1.83 V. These values are in comparison to hydrogen, which is assign an oxidation potential of 0 V.

15. Pereira N, Amatucci GG, Whittingham MS, Hamlen, R (2015) Lithium-titanium disulfide rechargeable cell performance after 35 years of storage. Journal of Power Sources 280:18–22.

16. The Inorganic Chemistry Laboratory at Oxford University has been declared a National Chemical Landmark by the Royal Society of Chemistry commemorating the groundbreaking work of John B. Goodenough and his team of researchers. The plaque was erected November 30, 2010.

17. Mizushima K, Jones PC, Wiseman PJ, Goodenough JB (1980) Li_xCoO_2 (0<x<1): a new cathode material for batteries of high energy density. Materials Research Bulletin 15:783–789.

18. Padhi AK, Nanjundaswamy KS, Goodenough JB (1997) Phospho-olivines as positive-electrode materials for rechargeable lithium batteries. Journal of the Electrochemical Society 144:1188–1194.

19. Yoshino A, Jitsuchika K, Nakajima T (1985) Li-ion battery based on carbonaceous material. Japanese Patent 1,989,293.

20. Akira Yoshino quoted by the Royal Swedish Academy of Sciences.

21. Williard N, He W, Hendricks C, Pecht M (2013) Lessons learned from the 787 Dreamliner issue on lithium-ion battery reliability. Energies 6:4682–4695.

22. International Energy Agency (2020) Global EV Outlook 2020. https://iea.blob.core.windows.net/assets/af46e012-18c2-44d6-becd-bad21fa844fd/Global_EV_Outlook_2020.pdf.

23. Cherif R, Hasanov F, Pande, A (2017) Riding the energy transition: oil beyond 2040. IMF Working Paper WP/17/120.

24. The actual mineral spinel is $MgAl_2O_4$. Many important oxides have this structure including the naturally occurring magnetic mineral, magnetite (Fe_3O_4), and other commercially useful magnetic materials. The Black Prince's ruby, the large red gem set in the Maltese cross in the front of the

Imperial State Crown of the United Kingdom is not a ruby at all, it is one of the world's largest gem-quality red spinels.

25. Fortier SM, Nassar NT, Lederer GW, Brainard J, Gambogi J, McCullough, EA (2018) Draft critical mineral list—summary of methodology and background information—U.S. Geological Survey technical input document on response to Secretarial Order No. 3359. U.S. Geological Survey Open-File Report 2018-1021, 15 p.

26. Li W, Erickson EM, Manthiram A (2020) High-nickel layered oxide cathodes for lithium-based automotive batteries. Nature Energy 5:26–34.

27. Winter M, Besenhard JO (1999) Electrochemical lithiation of tin and tin-based intermetallics and composites. Electrochimica Acta 45:31.

28. Mackay DT, Janish MT, Sahaym U, Kotula PG, Jungjohann KL, Carter CB, Norton MG (2014) Template-free electrochemical synthesis of tin nanostructures. Journal of Materials Science 49:1476–1483.

29. Barume B, Naeher U, Ruppen D, Schütte P (2016) Conflict minerals (3TG): mining production, applications and recycling. Current Opinion in Green and Sustainable Chemistry 1:8–12.

30. Data from the United States Geological Survey (USGS), available at https://pubs.usgs.gov/periodicals/mcs2021/mcs2021-tin.pdf.

31. Chan CK, Peng H, Liu G, McIlwrath K et al (2008) High-performance lithium battery anodes using silicon nanowires. Nature Nanotechnology 3:31–35.

32. Wagner RS, Ellis WC (1964) Vapor-liquid-solid mechanism of single crystal growth. Applied Physics Letters 4:89–90. The VLS process was developed by Wagner and Ellis, both at Bell Laboratories in Murray Hill, New Jersey, to grow silicon "whiskers" as small as 100 nm in diameter. (Nowadays we would refer to these as nanowires.) The process uses gold, which forms the "L" in VLS. The VLS mechanism has been used to grow many other types of nanowire.

33. https://amprius.com/2018/12/amprius-silicon-nanowire-lithium-ion-batteries-power-airbus-zephyr-s-haps-solar-aircraft.

34. Data from United States Geological Survey (USGS), available at https://pubs.usgs.gov/periodicals/mcs2021/mcs2021-diatomite.pdf.
    Diatomite is known by several other names, kieselguhr (Germany), tripolite (after an occurrence near Tripoli, Libya), and moler (an impure Danish form). Because U.S. diatomite occurrences are at or near the Earth's surface, recovery from most deposits is achieved through low-cost open pit mining.

35. Campbell B, Ionescu R, Tolchin M, Ahmed K et al (2016) Carbon-coated, diatomite-derived nanosilicon as a high rate capable Li-ion battery anode. Scientific Reports 6:33050.

36. Whittingham MS (2020) Special editorial perspective: beyond Li-ion battery chemistry. Chemical Reviews 120:6328–6330.

37. Specific energy density is the energy per weight. Ideally a battery for mobile applications would have a very high value of specific energy density. The

other parameter frequently used to compare different battery chemistries is volumetric energy density in units of Wh/l. Except for stationary applications more compact batteries are desirable.

38. Abraham KM (2020) How comparable are sodium-ion batteries to lithium-ion counterparts? ACS Energy Letters 5:3544–3547.

39. Bradley DC, Stillings LL, Jaskula BW, Munk LA, McCauley AD (2017) Lithium, chap. K of Schulz KJ, DeYoung JH, Jr, Seal RR, II, Bradley DC (eds) Critical mineral resources of the United States—Economic and environmental geology and prospects for future supply. U.S. Geological Survey Professional Paper 1802, pp K1–K21.

40. CRC handbook of chemistry and physics, 97th edn (2016–2017), pp 14–17.

41. United States Geological Survey (USGS), available at https://pubs.usgs.gov/periodicals/mcs2021/mcs2021-lithium.pdf. The data does not include the United States production to avoid disclosing company proprietary data. The only lithium production in the United States in 2020 was a brine operation in Nevada.

42. The values used here refer to the metal. An alternative way of reporting lithium production is in units of LCE or lithium carbonate equivalent. Values reported in LCE will always be higher than values giving just the metals weight. Lithium carbonate has the molecular formula $LiCO_3$. Weights reported in LCE will include the masses of carbon and oxygen, which are heavier atoms than lithium. To convert from lithium metal to LCE multiply by 5.323. So global reserves of 86 million tons of lithium would 458 million tons LCE. In either case, a very large amount.

43. Lithium-ion batteries for electric vehicles have a lifespan of 8–10 years or 100,000 miles or more.

44. Mayyas A, Steward D, Mann M (2019) The case for recycling; overview and challenges in the material supply chain for automotive lithium-ion batteries. Sustainable Materials Technology 19:e00087.

45. Sonoc A, Jeswiet J, Soo VK (2015) Opportunities to improve recycling of automotive lithium ion batteries. Procedia CIRP 29:752–757.

46. The Royal Society of Chemistry Periodic Table, available at https://www.rsc.org/periodic-table/element/27/cobalt.

47. United States Geological Survey Mineral Commodity Summary, available at https://pubs.usgs.gov/periodicals/mcs2021/mcs2021-cobalt.pdf.

48. Fu X, Beatty DN, Gaustad GG, Ceder G et al (2020) Perspectives on cobalt supply through 2030 in the face of changing demand. Environmental Science and Technology 54:2985–2993.

49. Gurney JJ, Levinson AA, Smith HS (1991) Marine mining of diamonds off the west coast of southern Africa. Gems and Gemology 27:206–219.

50. Heffernan O (2019) Deep-sea dilemma. Nature 571:465–469.

7

Here Comes the Sun

One undeniable reason for supporting the development of solar cell technology is the enormous amount of energy that comes from the Sun each and every day. Every *year* we use a total of 410 quintillion joules of energy. Every *hour* we receive 430 quintillion joules of energy from the Sun.[1] So, we get more energy from the Sun in 1 h than the world's population uses in an entire year. A pretty compelling statistic!

In 2020, about 4000 billion kilowatt-hours of electricity were generated at utility-scale electricity generation power plants in the United States. Over 60% of this electricity was from fossil fuels, mainly by burning natural gas and coal. About 20% was from nuclear energy and just under 20% was from renewable energy sources. A total of only 2.3%—91 billion kilowatt-hours—was from solar energy.[2] Most of that electricity was generated in California. Despite that very small percentage and with their overwhelming reliance on fossil fuels, it is only China that generates more electrical energy from the Sun than the United States with a current capacity to produce 261 terawatt-hours or 0.9 quintillion joules (0.9×10^{18} J).

Current solar technologies, the devices that convert sunlight into electrical energy, operate in one of two ways either through the photovoltaic (PV) effect or concentrating solar-thermal power (CSP). Solar cells using the photovoltaic effect account for well over 95% of the total amount of all solar generated electricity. In this chapter the focus is on the history of photovoltaic materials, approaches that have been used to increase efficiency and lower the cost of solar energy conversion, and possible directions for

© The Author(s), under exclusive license to Springer Nature
Switzerland AG 2023
M. G. Norton, *A Modern History of Materials*,
https://doi.org/10.1007/978-3-031-23990-8_7

the future. Projections by the United States Energy Information Administration (EIA) indicate that by 2050, domestic electricity production from solar will exceed 900 billion kilowatt-hours—a tenfold increase over the next three decades.[3] Whilst the majority of this electricity, as it does now, will likely come from photovoltaic installations, concentrating solar-thermal power systems may have an important future role if demand for solar energy outpaces photovoltaic capacity.

Using specially designed mirrors concentrating solar-thermal power systems reflect and concentrate sunlight onto receivers containing a high temperature fluid—typically a liquid mixture of sodium and potassium nitrates (mixed alkali metal salts). Heat generated in the liquid salts, which can reach a temperature of 565°C (1049°F), is used to create steam that spins a turbine that drives a generator to produce electricity. Excess heat can be used in a variety of industrial applications, for instance in seawater desalination as demonstrated by scientists from Golden, Colorado's National Renewable Energy Laboratory (NREL).[4]

If water rather than the salt mix is used as the fluid, steam can be generated directly by absorption of solar radiation then sent directly to the turbine. Using water/steam as the heat-transfer fluid has the benefit of eliminating the need for costly heat exchangers. Concentrating solar-thermal power technology is generally used for utility-scale plants with at least 1 megawatt of total electricity generating capacity.

The Ivanpah Solar Electric Generating System (ISEGAS) is the largest concentrating solar-thermal power plant in the world. Located in one of the hottest places on the planet—California's Mojave Desert—the plant is capable of producing 392 megawatts of electricity using 173,500 heliostats, each with two sun-tracking mirrors that focus sunlight onto three solar power towers. ISEGAS serves the electricity needs of more than 140,000 homes each year.

Spain, a country frequently associated with sunshine and warm weather, has several concentrating solar-thermal power systems. Planta Solar 10 (PS10) and Planta Solar 20 (PS20) both located west of Seville are water/steam systems with capacities of 11 and 20 megawatts, respectively. These were the first two operational solar-thermal facilities in the world. PS10 began commercial operation in 2007, PS20 came online two years later. Both facilities remain operational.

Located in the city limits of Fuentes de Andalucía about 30 miles east of Seville, the Gemasolar Thermosolar Plant produces nearly 20 megawatts of electricity using concentrating solar-thermal power. This is enough power to meet the needs of 27,500 households. The photograph in Fig. 7.1 shows

Fig. 7.1 The Gemasolar concentrating solar-thermal power plan in Fuentes de Andalucía, Seville, Spain. The heliostats set up around the tower reflect and concentrate solar energy onto the molten salt receiver located at the top of the tower (Reprinted from https://www.energy.sener/projects/gemasolar)

the 140-m high receiving tower surrounded by 310,000 square meters of mirror surface divided between 2650 heliostats. Gemasolar is operationally very different from both the Ivanpah Solar Electric Generating System and Planta Solar because it utilizes molten salt thermal storage. Brought on stream in 2011, Gemasolar is the first commercial plant in the world to use a combination of high temperature tower receiver technology together with molten salt thermal storage. The very high temperature of the molten salt mixture is used to store excess thermal energy produced during daylight hours providing sufficient energy to ensure the plant can remain operational for up to 15 h without sunlight. This extremely useful feature avoids the need for costly batteries or other forms of energy storage device frequently required by intermittent power sources including both solar and wind.

The United States Department of Energy has identified several advantages as well as challenges with concentrating solar-thermal power technology. Reliability and predictability are among its biggest advantages. There is essentially an unending supply of sunshine. On that we can rely. We know exactly when the Sun will rise and with equal precision when it will set. Because the "fuel" is free—unlike fossil fuels—the costs are predictable with over 60% of the total lifetime expenses occurring in the first year. But those costs, while known, are extremely high. Capital and maintenance is more expensive than other forms of power stations including solar photovoltaic plants. A 2019 study compared the costs of photovoltaic and concentrating solar-thermal power 100 megawatt plants in Saudi Arabia and found that capital costs for concentrating solar-thermal power were around $500 million compared to approximately $100 million for a comparable photovoltaic installation. The cost of electricity was also greater for concentrating solar-thermal power at $110 per megawatt hour compared to less than $40 per megawatt hour for the photovoltaic installation.[5] Despite these disadvantages, Avi Shultz Concentrating Solar-Thermal Program Manager in the Office of Energy Efficiency and Renewable Energy is optimistic about concentrating solar-thermal power technology: "Its flexibility and predictability will make it a strong contender for meeting our changing energy needs, today, tomorrow, and in 100 years."[6]

Converting sunlight into electrical energy using the photovoltaic effect is very different from the concentrating solar-thermal power approach, but conceptually is just as straightforward. Rather than the sunlight being captured by mirrors and reflected onto a receiver containing a fluid, the light instead is incident upon a photovoltaic panel comprising a number of individual cells each producing a small amount of power, about 1 or 2 watts. A semiconductor, nowadays most usually silicon, forms the active component of the cells, which when exposed to sunlight absorbs part of the incident solar energy.[7] It is this extra boost of energy that allows the electrons to break free from the bonds holding the atoms in the semiconductor together and flow through the material as an electric current. The current is extracted through conductive metal contacts—electrodes—placed on the top and bottom of the cell.

French physicist Alexandre-Edmond Becquerel was just nineteen years old when he discovered the photovoltaic effect in 1839.[8] Becquerel was experimenting in his father's laboratory with an electrolytic cell consisting of platinum and silver plates separated by a thin membrane immersed in an acidic silver chloride solution. When the electrodes were illuminated with sunlight, Becquerel noticed a significant generation of electricity. This experiment was the first demonstration of what became known as the photovoltaic effect, the direct conversion of light into electricity, which early on was also known as the "Becquerel effect". During his experiments, Becquerel also discovered the influence of the wavelength of visible light on the photovoltaic effect—blue light yielded the largest current generation.

In 1873 English electrical engineer Willoughby Smith discovered that the conductivity of the semiconducting element selenium, atomic symbol Se, significantly increased when exposed to an intense light source. The conductivity varied according to the amount of light that reached the material increasing from 15 to as much as 100%.[9] This property is known as photoconductivity—the increase in electrical conductivity of a material when it is exposed to light. Three years after Smith's discovery of photoconductivity in selenium two British scientists William Grylls Adams and his student Richard Evans Day discovered that the element also demonstrated the photovoltaic effect. Although the early selenium cells produced no useable power—their efficiency was only ½%—they did prove that a semiconducting material could be used to convert light into electricity without heat or moving parts.

American inventor Charles Edger Fritts made the first working solar cell in 1883. This device consisted of a wafer of selenium that was between 25 and 125 μm in thickness sandwiched between a brass plate on the bottom and on top a gold leaf so thin it was almost transparent. The metals acted as electrodes on either side of the semiconductor to extract the current generated by the incident sunlight. Fritts's cell worked but achieved an energy conversion rate of barely 1%. This was little better than Adams and Day's selenium cell and extremely low by today's standards. Fritts estimated the cost of a completed cell including all the materials at $100.[10] This would be equivalent to about $2600 in today's money. Fritts's results must have been encouraging enough for an array of the selenium cells to be installed on a New York City rooftop in 1884: The world's first rooftop solar array. A photograph from the time reproduced in Fig. 7.2 shows the installation and the wires that carried the solar generated electricity.

But it would be another 70 years before the full potential of photovoltaic technology would be realized.

In 1941 Bell Labs engineer Russell Ohl filed a United States patent for "Light-sensitive electric devices and more particularly to photo-E.M.F. [photovoltaic] cells comprising fused silicon of high purity."[11] Ohl described how the cell, a silicon p–n junction, converts light energy directly into electrical energy. He also noted the response of his device to different wavelengths of light. The electrons "are very efficiently released" when irradiated with infra-red or visible light and the device "shows some response for ultra-violet radiation." The importance of this observation would be realized later in the design of modern solar cells that attempt to optimize the choice, or combinations, of materials to maximize the range of solar wavelengths that are absorbed.

Using large p-n junctions formed by diffusing boron atoms into n-type silicon wafers Calvin Fuller, Daryl Chapin, and Gerald Pearson made the first silicon solar cells at Bell Labs in 1954. Importantly, these were the first solar

Fig. 7.2 Charles Fritts installed the first solar panels on this New York City rooftop in 1884 (Courtesy of John Perlin)

cells capable of converting enough of the Sun's energy into usable amounts of electrical power. This was demonstrated by using solar generated electricity to run a 21-inch Ferris wheel. The following day, April 26, 1954, *The New York Times* was particularly enthusiastic about the potential of solar cells, remarking on page one of the newspaper: "The silicon solar cell may mark the beginning of a new era, leading eventually to the realization of mankind's most cherished dreams – the harnessing of the almost limitless energy of the sun for the uses of civilization." An article in *U.S. News & World Report* titled "Fuel Unlimited" speculated that one day the silicon strips "may provide more power than all the world's coal, oil, and uranium." This speculation is presently far from being realized. Currently, photovoltaics account for only about 3% of global electricity generation.

Those early silicon solar cells made by Fuller, Chapin, and Pearson at Bell Labs had an efficiency of 2.3%, a significant improvement over the selenium cells. To demonstrate that solar cells could be a viable power source Daryl Chapin determined that a realistic efficiency goal was 5.7%. By the end of the 1950s the efficiency of silicon solar cells was over 10% and they were demonstrating their usefulness providing power for ground-based applications including rural telephone systems and extraterrestrial applications such as space satellites.

On August 7, 1959, the Explorer 6 satellite was launched with a photovoltaic array of 9600 cells (1 cm to 2 cm each). The small satellite's mission included studying galactic cosmic rays and radio wave propagation in the upper atmosphere. One week after launching Explorer 6 took the first image of Earth ever by a satellite. The very blurry image was captured over Mexico at an altitude of approximately 27,000 km.

The following decades saw further developments in the use of silicon solar cells including a Japanese lighthouse, a French solar furnace, a solar-powered aircraft—the Solar Challenger—that crossed the English Channel, a solar-powered car—the Quiet Achiever—that covered the almost 2800 miles between Sydney and Perth, and NASA's Orbiting Astronomical Observatory, powered by a one-kilowatt photovoltaic array. There were also developments in the materials used to make solar cells. In 1976 David Carlson and Christopher Wronski at the RCA David Sarnoff Research Laboratory in Princeton, New Jersey fabricated an amorphous silicon photovoltaic cell with a 5.5% efficiency.[12] Up to that time silicon solar cells had all used the crystalline form of the material. Amorphous silicon offered the significant possibility of reducing the cost of the device. Furthermore, the material was compatible with many existing thin film processing methods, which provided an added level of manufacturing flexibility when compared to crystalline silicon.

At the University of Delaware, a thin film solar cell using a combination of copper sulfide, Cu_2S, and cadmium sulfide, CdS, was made demonstrating an efficiency of more than 10%.[13] And in 1992 researchers at the University of South Florida produced a thin-film photovoltaic cell made of cadmium telluride, $CdTe$, that had an efficiency of 15.9%.[14] Two years later, researchers at the National Renewable Energy Laboratory fabricated a solar cell made from the compound semiconductors gallium indium phosphide, GaInP, and gallium arsenide, GaAs, that exceeded 30% conversion efficiency. In 2020, the National Renewable Energy Laboratory beat out all previous efficiency records including their own with an engineered six-junction cell having a remarkable efficiency of 47.1% under concentrated illumination equivalent to 143 Suns![15] Each junction was precisely engineered to capture a specific range of solar wavelengths. Even under single Sun conditions cell efficiency was still an impressive 39.2%.

Efficiency is an essential performance metric because it is a measure of how effective a device is at converting energy from one form into another. In the case of photovoltaic cells efficiency is defined as the amount of electrical energy coming out of the cell compared to the energy from the light shining on it. Theoretically the limit for conversion of radiative energy into electrical energy can be as high as 90%. Depending on the type of absorber technology and the intensity and wavelength of the incoming light actual practical efficiencies will be much lower than theory predicts.

The most important electronic property of solar absorbers is the bandgap energy, which determines the possible wavelengths of light that a material can absorb and convert into an electric current. For silicon, the bandgap energy is 1.12 electron volts, in the infra-red region of the electromagnetic spectrum. Parts of the solar spectrum with energy less than 1.12 electron volts (eV) cannot be absorbed by silicon to create the free electrons that carry current. This energy is absorbed only as heat—the material will warm up. Solar radiation with energies above the bandgap energy will also be transformed into heat, which lowers cell efficiency because the heat cannot be utilized in the photovoltaic process. These limitations on either side of the bandgap energy implies that ¼ of the solar energy reaching the Earth cannot be converted into electricity by a silicon semiconductor alone. Of the many free electrons created by the incident radiation not all of them will make it to the metal contacts to form an electric current. Because of these reasons the theoretical efficiency of silicon photovoltaic cells is limited to about 33%.[16]

Despite not having the highest conversion efficiency silicon remains by far the most widely used absorber material for solar cells, because it provides an important combination of low cost and long service life. Over 80% of present-day photovoltaic panels are based on high purity silicon. The excellent stability of silicon to weathering guarantees that installed photovoltaic modules last for twenty-five years or more, while still producing at least 80% of their original power after that time. About 12% of all the silicon produced worldwide goes into the manufacture of solar cells. That is four times more than is used to make silicon chips for the microelectronics industry.

Researchers in industry, national laboratories, and universities continue to explore alternatives to silicon with the goals of reducing cost, decreasing the amount of absorbing material required, while increasing efficiency.[17] Some of these approaches have met with more success than others, but none has yet to make a considerable dent in the market dominance of silicon photovoltaics.

The second most common photovoltaic material after silicon is the compound semiconductor cadmium telluride, which represents about 5% of the world market. Rather than being in the form of crystal wafers, cadmium telluride solar cells use a thin film of the photovoltaic material deposited on a supporting substrate typically aluminum and protected by a transparent glass layer. Manufacturing thin films is much less expensive, and quicker, than the process used to make silicon single crystals, which was described in Chapter 4. These advantages make cadmium telluride solar cells a cost-effective alternative to crystalline silicon.[18] Early cadmium telluride thin film cells made in the mid 1970s and early 1980s by Matsushita in Japan and Kodak in the United States demonstrated efficiencies less than 10%, well below those for the crystalline silicon devices being made at that time by Mobil Solar, RCA, and Sandia National Laboratories. Over time cell efficiencies have increased. The present record for a laboratory tested cadmium telluride solar cell is held by American manufacturer First Solar at 22.1%. Commercial modules have published efficiencies as high as 18.2% in 2021, which is close to the value for polycrystalline silicon cells.

Manufacturers of cadmium telluride solar cells are working on increasing cell efficiencies by improving the quality of their thin films. Defects in the films formed during growth can reduce an important property called the minority carrier lifetime, which means that free electrons created in the semiconductor do not contribute to the electrical current because they become trapped within the material. In the early days of the silicon transistor industry defects in the single crystal *boules* would cause device performance problems. The crystal growers were able to develop methods to produce defect-free crystals of increasing sizes, which became a major factor in the rapid expansion of the microelectronics industry. Improvements in the processing of cadmium telluride films with a reduction in the defect concentration will no doubt lead to increased efficiencies when the material is used as a solar absorber.

But cadmium telluride is not without its own disadvantages. Two specific issues faced by cadmium telluride cells are the toxicity of cadmium and the scarcity of tellurium. Cadmium is a very toxic heavy metal with a long biological half life (up to thirty years.)[19] Its link to many diseases including kidney disease, osteoporosis, and cancer has caused cadmium to be removed from a number of applications including as a red colorant in glass. But its use in photovoltaic cells has increased as these cells have demonstrated improved performance. Cadmium is largely obtained as a separated by-product of zinc mining. Zinc producers are obliged to remove it during refining. Cadmium can also be recovered by recycling spent nickel cadmium "NiCad" batteries. The same process can be used to recycle old cadmium telluride solar panels.

Tellurium is a by-product of copper mining and historically there was little incentive to extract it because there was such a limited market for this metal. With the growing deployment of cadmium telluride solar panels, the situation is now quite different. Even doubling of the supply of tellurium may not be enough to cover the predicted market demand for thin-film solar cells. The United States Department of Energy anticipates a supply shortfall as early as 2025, which stresses the need for more extensive recycling and reuse programs. Unfortunately, we won't have long to wait to see if this projection is right or wrong.

A close relative of cadmium telluride is copper indium gallium diselenide, known as CIGS. By tailoring the composition this compound semiconductor we can create a material that has almost optimal properties for capturing the Sun's energy. In the laboratory scientists at Solar Frontier have demonstrated efficiencies of CIGS photovoltaic cells of over 23%. But the complexity of controlling the combination of four different elements—copper, indium, gallium, and selenium—makes the transition from the laboratory to industrial-scale manufacturing challenging. A major downside of CIGS cells is that they are less resilient than silicon to the effect of weathering and the harsh outdoor environment in which solar cells have to operate.

Another type of thin-film cell, similar to cadmium telluride, is the perovskite solar cell, named after the material's characteristic crystal structure. The mineral perovskite refers to the compound calcium titanate, which has the chemical formula $CaTiO_3$. Perovskite was discovered in Russia's Ural Mountains in 1839 by German mineralogist Gustavus Rose and is named after Russian mineralogist and nobleman count Lev Aleksevich von Perovski.

The ideal perovskite structure is cubic and near-perfect cubes of calcium titanate can be found in nature. But for many compounds with this structure, including calcium titanate, it was recognized early on by crystallographers that their structure was not truly cubic, but rather a slight distortion from the perfect cube. Helen Dick Megaw a crystallographer working in the Material Research Laboratory, part of Philips Lamps Ltd, used the X-ray diffraction camera in Lawrence Bragg's laboratory at the Cavendish Laboratory in Cambridge to show that the structure of barium titanate, a compound closely related to the mineral perovskite, was simply derived from the cubic structure by stretching it along one axis by about one percent—a tiny distortion.[20] Megaw also showed that if the crystal of barium titanate was heated to 200 °C that its structure would indeed revert to that of a perfect cube. The temperature at which this transformation occurs became known as the Curie temperature.

While the history of cadmium telluride and CIGS photovoltaic cells dates from the mid 1970s the exploration of perovskite materials for solar energy capture happened less than a decade ago. The first perovskite solar cells were built and tested in 2013 at Swiss laboratory EPFL located in the beautiful city of Lausanne. These cells had an efficiency of 14%. Seven years later efficiencies greater than 25% had been achieved by scientists working at Ulsan National Institute of Science and Technology (UNIST) in South Korea. Among the many different solar absorbers that have the perovskite structure, compounds in the methylammonium lead trihalide, $CH_3NH_3PbX_3$, family are the most commonly studied. As these are trihalides, X can be a halogen either iodine, bromine, or chlorine. The chemical formula is considerably more complex than that of the mineral perovskite, but we can map both compounds to the perovskite-type structure with X replacing oxygen, lead replacing titanium, and the methylammonium cation ($CH_3NH_3^+$) occupying the site that calcium would take in the eponymous mineral.[21] The beauty of the perovskite structure is that there are many substitutions that can be made in the basic ABX_3 formula, which is one of the reasons that perovskites find numerous important electronic applications.

Combining perovskite absorbers with either conventional silicon or CIGS thin films creates what are called tandem cells. What is exciting about tandem cells is that by choosing two materials with different bandgap energies we expand the absorption range of the device allowing more of the Sun's abundant energy to be converted into electricity than would be possible in a single cell.

On the top of a tandem cell is the large bandgap material that absorbs the higher energy parts of the visible solar spectrum (those with energies above that of green light). The bottom part of the cell, made of the lower bandgap material absorbs the low energy radiation with energies in the red even into the long infra-red wavelengths. Tandem cells combining perovskites and silicon were developed by American solar panel manufacturer First Solar in 2018 with demonstrated efficiencies around 24%. Silicon, on the bottom of the cell, has a bandgap energy lower than that of the perovskite absorbers. As of 2021, the most recent versions of silicon perovskite tandem cells are made by Oxford PV, a spin-out company from the University of Oxford. These cells have efficiencies around 30%.

Perovskite/CIGS tandem cells were developed at the University of California Los Angeles in 2019. Using this material combination solar cells produced by Japanese photovoltaic company Solar Frontier have measured efficiencies of just over 23%. In the laboratory setting, perovskite solar cell efficiencies have improved faster than any other photovoltaic material. It is worth noting that these values are in the same ballpark as the best that is being obtained using single crystal silicon—a material that has been supported by over four decades of research and development. Despite the rapid improvement in efficiency, perovskite cells or tandem combinations are not yet commercially viable. For one thing, the material does not match the durability of silicon. Further work is necessary to produce a perovskite absorber material that can survive the required 20 years outdoor lifetime.

Two other types of material that are under investigation as solar absorbers are organic photovoltaics or OPVs and quantum dots. While neither technology has had any significant commercial impact the advantages of each suggest that they may become important in the future, possibly as part of a tandem cell.

Organic photovoltaic cells (often referred to as plastic solar cells) are composed of special types of conductive carbon-rich polymers. Like many organic compounds the compositions of organic photovoltaics can be tailored to enhance a specific function. For a particular material, the property of interest may be optimizing the bandgap energy to increase the region of solar wavelengths that are absorbed. Or maybe we want to make the material more mechanically flexible so that it can be applied onto a bendable plastic substrate. It is also possible to change the color of organic photovoltaic materials to enhance their aesthetic appeal while sacrificing absorption of some parts of the visible solar spectrum. Early work on plastic solar cells was led by the Johannes Kepler University of Linz in Austria during the early 2000s.[22] At the time cell efficiencies were less than 4%, but there was considerable excitement about the potential of this technology. Imagine low-cost plastic films cast between panes of glass forming power generating windows. Organic photovoltaic cells could convert these large unused areas—millions of square feet—into an integral component of sustainable low-carbon buildings and add to the limited roof top space available in crowded cities for placing solar panels.

Organic photovoltaic technology relies on polymers that conduct electricity. Our general view of plastics, at least up to the end of the 1970s, was that they are all very good electrical insulators. Engineers have used the insulating properties of plastics for decades to isolate electrical connections in applications from undersea telephone cables to on-board aircraft radar. But our perception of plastics was dramatically changed by three chemists Alan Heeger, Alan MacDiarmid, and Hideki Shirakawa, who demonstrated that it was possible to produce electrically conductive polymers. To do this required forming alternating single and double bonds between the carbon atoms that link together to comprise the long molecular chains that make up the polymer, then doping the material through the addition of suitable atoms. The effect of adding the dopant is that it generates free electrons or electron holes (positive charge carriers) that can move along the molecule creating an electrical current. Heeger, MacDiarmid, and Shirakawa overlapped at the University of Pennsylvania. Hideki Shirakawa recounts one of the critical experiments performed in November 1976 two months after he joined Penn at the invitation of MacDiarmid: "I still vividly remember the day … I was measuring the electric conductivity of polyacetylene [an organic polymer with the repeating unit $(C_2H_2)_n$ made by polymerizing ordinary acetylene welding gas] by the four-probe method, adding bromine. At exactly the moment we added bromine, the conductivity jumped so rapidly that he [postdoctoral fellow C.K. Chiang] couldn't switch the range of the electrometers. Actually, the conductivity was ten million times higher than before adding bromine. This day marked the first time we observed the doping effect, although it was a pity that the expensive equipment was broken."[23]

The importance of this work was recognized by the Royal Swedish Academy of Sciences with Heeger, MacDiarmid, and Shirakawa sharing the 2000 Nobel Prize in Chemistry "for the discovery and development of conductive polymers."[24] The press release accompanying the announcement of the award concluded: "In the future we will be able to produce transistors and other electronic components consisting of individual molecules—which will dramatically increase the speed and reduce the size of our computers. A computer corresponding to what we now carry around in our bags would suddenly fit inside a watch … "[25] The predication in the last sentence proved correct, but it was not made possible by molecular electronics and conducting polymers, but rather the even greater improvements in silicon microelectronics.

The best plastic solar cells to date, which use the fullerene C_{60} buckyball molecule as an additive, made by a group of researchers led by Shanghai Jiao Tong University in China have efficiencies barely greater than 18%.[26] Currently, these materials lack the stability and reliability of crystalline silicon, but in the future they could be considerably less expensive to manufacture in high volumes. Just like many other plastics organic photovoltaic materials are processed at temperatures close to room temperature, which means they can be applied to a variety of supports including flexible materials.

A rapidly growing area of materials research for solar cells is quantum dots (QDs). These are tiny semiconductor particles—just a few nanometers across—that have optical and electronic properties that differ from larger particles due to the peculiar effects of quantum mechanics. One of the first quantum dot solar absorbers that was successfully made into a solar cell is lead sulfide, PbS. The left-hand side image shown in Fig. 7.3 is of a single layer of lead sulfide quantum dots synthesized at the Massachusetts Institute of Technology.[27] What is remarkable is that if more quantum dots are deposited, they will self-assemble into ordered three-dimensional superlattice structures as shown in the middle image in Fig. 7.3. The researchers at MIT estimate that the superlattice is five quantum dots thick. The right-hand image is an

Fig. 7.3 Lead sulfide quantum dots. **a** The transmission electron microscope image of a quantum dot monolayer shows the self-assembly into a hexagonally close packed arrangement. **b** The transmission electron microscope image shows a three-dimensional superlattice approximately 5 layers thick. The X-ray pattern in **c** shows the ordering of the superlattice into a body-centered cubic arrangement (Reprinted from Weidman, Mark C., Beck, Megan E., Hoffman, Rachel S., Prins,Ferry, & Tisdale, William A. (2014) Monodisperse, air-stable PbS nanocrystals via precursor stoichiometry control. ACS Nano 8:6363-6371. https://pubs.acs.org/doi/10.1021/nn5 018654. Further permissions related to the material excerpted should be directed to the ACS)

X-ray diffraction pattern that confirms ordering of the quantum dots into a cubic structure.

Early quantum dot solar cells had efficiencies similar to those of some of the first plastic photovoltaic cells—around 4%. It is very difficult to create good electrical connection between individual dots, which is one factor that limits device efficiency. But research in new coating materials and different materials combinations has led to increases in efficiency such that the best quantum dot solar cells are matching the performance of their plastic counterparts.

There are two characteristics of quantum dots that make them particularly exciting for solar cells. Firstly, their bandgap energy is customizable. It depends on the size of the dots. Doris Segets at the Institute of Particle Technology in Erlangen, Germany working with scientists at the University of California Berkeley and at Lawrence Berkeley National Laboratory showed that larger lead sulfide particles have smaller bandgap energies than smaller particles.[28] This size-dependent property of quantum dots allows them to be synthesized in such a size range that they can collect light that is difficult to capture by other materials that have a single fixed bandgap energy.

A second advantage of quantum dots is that they are, perhaps rather surprisingly, easy to make and they can be applied onto a substrate using simple techniques for instance spraying or roll-to-roll printing similar to the process used to print newspapers. The ease of processing allows quantum dot solar cells to be combined with other solar absorbers, for instance perovskites, to optimize the performance of a tandem solar cell. One can imagine a quantum dot tandem cell where each layer consists of nanoparticles of different sizes, with each layer absorbing a specific range of solar energies.

Another strategy to improve photovoltaic cell efficiency is layering multiple semiconductors to make what are called multijunction solar cells. These are similar to the tandem cells mentioned earlier, but they have more than two layers. Each layer can be selected to have a unique bandgap energy so they can absorb in specific parts of the solar spectrum, making greater use of the available sunlight than single-junction solar cells. Light that doesn't get absorbed by the first semiconductor layer is captured by the layer beneath it and so on. This is the approach used by researchers at the NREL who currently hold the world record efficiency for a solar cell using a six-junction solar cell, NREL 6-J. As you might imagine multijunction solar cells are very costly and difficult to manufacture so they are reserved typically for space exploration and military applications such as drones.

We can combine the concept of photovoltaics with concentrating solar-thermal power in what is termed concentrating photovoltaics or CPV. Instead of focusing sunlight on a receiver containing water or a high temperature fluid as in concentrating solar-thermal, light is focused onto a small photovoltaic panel using a mirror or a lens. By using an intense focused beam of sunlight less of the absorber material is required, which reduces costs. Also, the cells become more efficient as the light becomes increasingly concentrated, so the highest overall efficiencies are obtained with concentrating photovoltaics cells. However, the performance benefits of concentrating photovoltaic technology are primarily seen with expensive multijunction cells. The added costs of tracking the movement of the Sun make it difficult to compete commercially with today's high volume silicon modules.

While there have been considerable developments in new materials—from conducting plastics to quantum dots—for solar absorbers silicon solar cell manufacturers have made considerable progress in lowering processing costs. One example of this success is the fluidized bed reactor (FBR), which uses up to 90% less energy than conventional processes for synthesizing high purity silicon.

As mentioned earlier, most commercially available photovoltaic modules rely on crystalline silicon as the absorber material. These modules share a common process step in requiring a supply of polycrystalline (polysilicon), a fine-grained product that is usually in the shape of rods or beads depending on the method of production. Polysilicon is commonly manufactured from highly reactive gases synthesized from impure silicon (so-called "metallurgical-grade" silicon) obtained from quartz sand. In one process, named after the German company Siemens, a mixture of trichlorosilane ($SiHCl_3$) and hydrogen is passed over a heated silicon filament causing additional silicon atoms to deposit on the filament, which ultimately grows into a large upside-down U-shaped polysilicon rod. The released hydrogen and chlorine atoms are then reused in part of a closed loop process.

Because of the large amount of energy required it is a severe challenge for the Siemens process to continue to meet the scale and cost reduction requirements of the solar industry.

The fluidized bed reactor process for production of polysilicon was developed, at least in part, because of the demand from the solar cell industry for a lower cost and less energy intensive product.[29,30] Fluidized bed reactors produce granular silicon by decomposition of silane (SiH_4) gas. The process begins with silicon seed particles ranging from 100 to 2000 μm in diameter that are fluidized with a carrier gas such as hydrogen. An important and significant aspect of the fluidized bed reactor is the temperature range over which silicon can be produced. Gas decomposition can be achieved at temperatures as low as 600°C, as opposed to temperatures up to 1200°C that are used to produce bulk polysilicon via the Siemens process.

Once a desired particle size range has been achieved within the reactor the granules fall to the bottom where they are collected without disrupting the growth of the rest of the material. The continuous nature of the fluidized bed process is another benefit compared to the batch-style Siemens approach.

Figure 7.4 shows an etched cross-section sample of a silicon granule collected from a fluidized bed reactor operated by Renewable Energy Corporation (REC) in Moses Lake, Washington. Immediately apparent in the image is the series of concentric rings, "growth rings", caused by porosity differences within the granule as it moves between hotter and colder regions of the vertical reactor. Existing fluidized bed reactors can generate over 13,500 tons of polysilicon annually.

Solar photovoltaic capacity has doubled every two years over the past two decades. In the last ten years alone, the United States has seen almost a 50% average annual growth rate in installed solar capacity. There are now more than 81 gigawatts of solar capacity installed nationwide, enough to power 15.7 million homes. Worldwide cumulative installation capacity at the end of 2019 exceeded 580 gigawatts.[31] At the same time as capacity has increased, the cost of silicon photovoltaic cells has significantly decreased dropping by more than 70% over the last decade.[32] Perhaps not surprisingly crystalline silicon solar cells continue to dominate the global photovoltaic market with about a 95% market share and offer a highest cell efficiency of

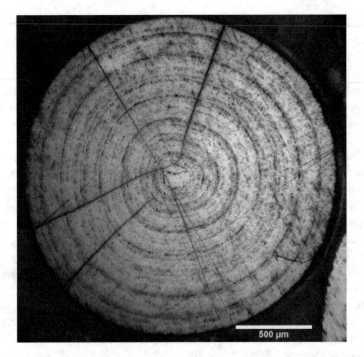

Fig. 7.4 Optical micrograph of an etched silicon granule from a fluidized bed reactor. The microstructure of the bead was revealed by etching with a mixed acid solution composed of hydrofluoric acid, nitric acid, and acetic acid (Reprinted from Dahl MM, Bellou A, Bahr DF, Norton MG, Osborne EW (2009) Microstructure and grain growth of polycrystalline silicon grown in fluidized bed reactors. Journal of Crystal Growth 311:1496–1500 with permission from Elsevier)

26.6% for single crystalline cells and 22.3% for multi-crystalline wafer-based technology.[33,34]

The importance of silicon was highlighted in Chapter 4 as the material behind the eponymous "chip". It is also the leading commercial solar cell material. In the next chapter we will see how silicon may be the dominant material for quantum computers.

Notes

1. A quintillion is a number followed by 18 zeros. So, 1 quintillion Joules is 1,000,000,000,000,000,000 J. Using decade power notation we can write 1 quintillion as 1×10^{18}.
2. U.S. Energy Information Administration. https://www.eia.gov/tools/faqs/faq.php?id=427&t=3.

3. U.S. Energy Information Administration. https://www.eia.gov/todayinen ergy/detail.php?id=42655.
4. Sharan P, Neises T, Turchi C (2018) Optimal feed flow sequence for multi-effect distillation system integrated with supercritical carbon dioxide Brayton cycle for seawater desalination. Journal of Cleaner Production 196:889–901.
5. Awan AB, Zubair M, Praveen RP, Bhatti AR (2019) Design and comparative analysis of photovoltaic and parabolic trough based CSP plants. Solar Energy 183:551–565.
6. United States Department of Energy, Office of Energy Efficiency and Renewable Energy. (2017). https://www.energy.gov/eere/articles/concentra ting-solar-power-could-provide-flexibility-and-reliability-us-electric-grid.
7. The solar spectrum covers a wide range of wavelengths from high energy gamma rays to the lower energy radio waves. The peak in intensity occurs in the visible part of the spectrum (around blue). It is in the visible range and tailing into the infra-red where most PV solar cells operate. Ideally as much of the solar spectrum should be captured. CSP systems use infra-red (heat) radiation.
8. Becquerel AE (1839) Recherches sur les effets de al radiation chimique de la lumiere solaire. Comptes Rendus l'Académie des Sciences 9:561–567.
9. Smith W (1873) Effect of light on selenium during the passage of an electric current. Nature 7:303.
10. Fritts CE (1883) On a new form of selenium cell, and some electrical discoveries made by its use. American Journal of Science 26:465.
11. Ohl RS (1946) Light-sensitive electric device. U.S. Patent 2,402,662. The patent application was filed on May 27, 1941 and was issued June 25, 1946.
12. Carlson DE, Wronski CR (1976) Amorphous silicon solar cell. Applied Physics Letters 28:671. This paper has been cited almost 2000 times.
13. Barnett AM, Meakin JD, Rothwarf A (1978) Progress in the development of high efficiency thin film cadmium sulfide solar cells. In: Photovoltaic Solar Energy Conference, Luxembourg, September 27–30, Proceedings. (A78-52776 24-44) Dordrecht, D. Reidel Publishing Co. pp 535–546.
14. Chu TL, Chu SS, Ferekides C, Wu CQ, Britt J, Wang C (1992) High efficiency thin film CdS/CdTe heterojunction solar cells. Journal of Crystal Growth 117:1073.
15. Geisz JF, France RM, Schulte KL, Steiner MA, Norman AG, Guthrey HL, Young MR, Song T, Moriarty T (2020) Six-junction III–V solar cells with 47.1% conversion efficiency under 143 Suns concentration. Nature Energy 5:326–335.
16. Shockley W, Queisser HJ (1961) Detailed balance limit of efficiency of p–n junction solar cells. Journal of Applied Physics 32:510.
17. Nikos Kopidakis at the National Renewable Energy Laboratory (NREL) maintains an excellent timeline, which is regularly updated, showing the history of solar cell materials. The efficiency data and chronologies mentioned in this chapter all come from this extremely useful publicly

available source: https://upload.wikimedia.org/wikipedia/commons/a/aa/Cel lPVeff%28rev210104%29.png.

18. In 2009, solar cells made of cadmium telluride became the first to undercut bulky silicon panels in cost per watt of electricity generating capacity ($/W). High quality single crystal silicon is made in bulk quantities using the Czochralski process named after Jan Czochralski. The history of this revolutionary process is described in: Uecker R (2014) The historical development of the Czochralski method. Journal of Crystal Growth 401:7–24.

19. Rahimzadeh MR, Rahimzadeh MR, Kazemi S, Moghadamnia A (2017) Cadmium toxicity and treatment: an update. Caspian Journal of Internal Medicine 8:135–145.

20. Megaw HD (1945) Crystal structure of barium titanate. Nature 155:484–485.

21. Eames C, Frost JM, Barnes PRF, O'Regan BC, Walsh A, Islam MS (2015) Ionic transport in hybrid lead iodide perovskite solar cells. Nature Communications 6:7497.

22. Brabec CJ, Sariciftci NS, Hummelen JC (2001). Plastic solar cells. Advanced Functional Materials 111:15–26.

23. Ideki Shirakawa – Nobel Lecture. NobelPrize.org. Nobel Prize Outreach AB 2022. Sun. 9 Jan 2022. https://www.nobelprize.org/prizes/chemistry/2000/shirakawa/lecture/.

24. The Nobel Prize in Chemistry 2000. NobelPrize.org. Nobel Prize Outreach AB 2021. Sun. 1 Aug 2021. https://www.nobelprize.org/prizes/chemistry/2000/summary/.

25. Press release. NobelPrize.org. Nobel Prize Outreach AB 2021. Sat. 31 Jul 2021. https://www.nobelprize.org/prizes/chemistry/2000/press-release/.

26. Zhang M, Zhu L, Zhou G et al (2021) Single-layered organic photovoltaics with double cascading charge transport pathways: 18% efficiencies. Nature Communications 12:309.

27. Weidman MC, Beck ME, Hoffman RS, Prins F, Tisdale WA (2014) Monodisperse, air-stable PbS nanocrystals via precursor stoichiometry control. ACS Nano 8:6363–6371.

28. Segets D, Lucas JM, Klupp Taylor RN, Scheele M et al. (2012) Determination of the quantum dot band gap dependence on particle size from optical absorbance and transmission electron microscopy measurements. ACS Nano 6:9021–9032.

29. Dahl MM, Bellou A, Bahr DF, Norton MG, Osborne EW (2009) Microstructure and grain growth of polycrystalline silicon grown in fluidized bed reactors. Journal of Crystal Growth 311:1496–1500.

30. Osborne EW, Spangler MV, Allen LC, Geertsen RJ, Ege PE, Stupin WJ, Zeininger G (2011) Fluid bed reactor. United States Patent 8,075,692 B2.

31. Photovoltaics Report, Fraunhofer Institute for Solar Energy Systems with support of PSE Projects GmbH (2020) Available from: https://www.ise.fraunhofer.de/content/dam/ise/de/documents/publications/studies/Photovoltaics-Report.pdf.

32. Solar Energy Industries Association (SEIA). Available from: https://www.seia.org/solar-industry-research-data.

33. Wang Z, Zhu X, Zuo, S, Chen M, Zhang C, Wang C, Ren X, Yang Z, Liu Z, Xu X, Chang Q, Yang S, Meng F, Liu Z, Yuan N, Ding H, Liu S, Yang D (2020) 27%-efficiency four-terminal perovskite/silicon tandem solar cells by sandwiched gold nanomesh. Advanced Functional Materials 30:1908298.

34. Green MA, Dunlop ED, Hohl-Ebinger J, Yoshita M, Kopidakis N, Ho-Baillie AWY (2020) Solar cell efficiency tables (Version 55). Progress in Photovoltaics: Research and Applications 28:3–15.

8

Certain About Uncertainty

One hundred years ago the physics that would be necessary to understand how quantum computers work, much less how to build them, did not exist. In the early 1920s electrons were viewed as tiny negatively charged particles—microscopically small spheres—with energies and locations within an atom that were well defined. In Niels Bohr's Nobel-prize-winning atomic model, electrons moved in orbits around the nucleus with their energy and position both known with, to many scientists, reassuring certainty.

Certainty is an essential feature of the operation of most current computers. In a conventional laptop or desk top computer or even today's massively powerful supercomputers, information is stored in *bits* that are defined as representing either a 0 or a 1. These binary states correspond to the location of electrons within each transistor that forms part of an integrated circuit. Figure 8.1 illustrates the most important component of an integrated circuit, a field effect transistor (FET) or, in this case, more specifically a metal oxide semiconductor (MOS) FET. The original United States patent for a field effect transistor was filed in October 1926 by Polish American physicist and electrical engineer Julius Edgar Lilienfeld. He had filed in Canada one year earlier, so the priority date for the field effect transistor is 1925.[1] This was twenty-two years before the Bell Labs trio of John Bardeen, Walter Brattain, and William Shockley invented the first working transistor– the point-contact transistor. In Lilienfeld's invention he proposed to use copper sulfide as the semiconducting material. At that time, the semiconductor effect was yet to be demonstrated in the elemental semiconductors, silicon and germanium.

© The Author(s), under exclusive license to Springer Nature
Switzerland AG 2023
M. G. Norton, *A Modern History of Materials*,
https://doi.org/10.1007/978-3-031-23990-8_8

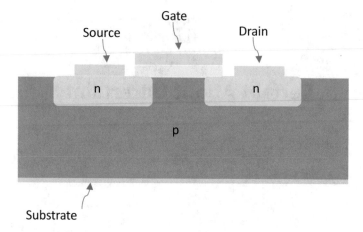

Fig. 8.1 Cross section diagram of a metal oxide semiconductor field effect transistor or MOSFET. The grey regions are *n*-type silicon, the red region is *p*-type silicon. As drawn, there is no conducting path between the Source and the Drain. The yellow region is the gate oxide—a thin insulating region

In 1933 Lilienfeld was awarded a further patent describing a MOSFET very similar to that shown in Fig. 8.1. There is no evidence, however, that he ever constructed a working version of either device.

Each MOSFET, just like the graphene field effect transistor shown earlier in Fig. 5.4, consists of three essential components—a source, a gate, and a drain. In the example shown in Fig. 8.1, both the source and the drain are doped *n*-type, meaning they have an excess of electrons. Separating the source and drain is a region of *p*-type silicon where there is an excess of positively charged electron holes. As drawn in Fig. 8.1, there is no conducting pathway between the source and the drain because an electrically resistive barrier exists at each *p*–*n* junction. This barrier is called the contact potential. Key to the operation of a MOSFET is the gate—a metal separated from the underlying *p*-type semiconductor by a very thin insulating oxide layer.[2] The path between the source and drain is either open or closed, depending upon the voltage applied to the gate electrode. There is no middle ground.

If a positive voltage is applied to the gate, electrons (which exist as minority charge carriers in the *p*-type material; electron holes are the majority charge carriers) move toward the gate forming a conductive path between source and drain. All the information—our photographs, emails, bank statements, etc.—stored in the computer is described by combinations of the two binary numbers, 0 and 1, corresponding to whether, or not, there is current flow between the source and drain. The enormous storage capacity and the extraordinary speed at which computer operations can be performed is possible because of the integration of billions of closely spaced transistors on a single silicon chip. Although Fig. 8.1 is not drawn to scale, a typical gate length for silicon transistors is about 10 nm.

Quantum computers, a technology unknown to most people a decade ago or so, lack the certainty of conventional computers relying instead for their operation on what many scientists say is the "weirdness" that happens at dimensions on the atomic scale and below. There are two quantum mechanical properties that are particularly useful for next generation computing technology. These are superposition and entanglement. The principle of superposition is that a *qubit*—the quantum computer analog of a *bit*—can be in multiple states at once. In other words, an individual qubit could be representing a 0 or a 1 or anywhere in between all at the same time. This is certainly very weird behavior when viewed through the eyes of classical physics. The science behind qubits is extremely complicated, particularly if we try to compare it to traditional bits.

Entanglement is the other property that is essential in the operation of a quantum computer. Entanglement is when qubits interact, working cooperatively with one another over distances much greater than the spacing between the atoms in a crystal. Again, this is very different from how, for instance, electrons behave inside a traditional computer chip. An example of entanglement happens in a class of materials called superconductors when they are cooled to just a few degrees above the absolute zero of temperature. Superconductors are described in more detail in the next chapter, but they provide here a useful example of entanglement. Once the material is in the superconducting state there is an attractive force between electrons, which in the right circumstances causes them to move in pairs (called Cooper pairs after their discoverer Leon Cooper). These electron pairs can interact over very large distances where the normal mechanisms of resistance to an electric current cannot operate. Superconductors experience zero electrical resistance.

It is these dual properties of superposition and entanglement that allow quantum computers, qubits, to process more information than conventional bits that can only be in a single state acting independently of each other. Using the analogy of lasers and lightbulbs, both are useful forms of light but very different each with their own applications and their own advantages and disadvantages, Thaddeus Ladd at California-based HRL Laboratories compared quantum computers with conventional computers this way: "Likewise, a quantum computer will not be a faster, bigger or smaller version of an ordinary computer. Rather, it will be a different kind of computer, engineered to control coherent quantum mechanical waves for different applications."[3]

The potential power of quantum computing was highlighted in 1994 when Bell Labs mathematician Peter Shor showed that a quantum computer could theoretically be used to solve certain mathematical problems that would be very difficult with a classical computer.[4] The result, which was presented at the Annual Symposium on Foundations of Computer Science in Santa Fe, New Mexico caused great excitement in the scientific community and became a powerful motivator for the design and construction of quantum computers. Unfortunately, at the time no one knew how to build even the most basic element of a quantum computer let alone make a full-size functioning machine. But, as Shor noted, " … a quantum computer … seems as though it could be possible within the laws of quantum mechanics."

In conventional computers the operation, creation of the binary 0 s and 1 s, is determined by where the electrons are in tiny regions of silicon just a few nanometers wide. In quantum computers we need to control not the position of the electron, but its spin. As first mentioned in Chapter 2, the concept of the spinning electron was a critical step in understanding atomic structure, how an atom's electrons fill the available energy states, and why atoms within a particular group of the Periodic Table of Elements have a similar set of properties. That discovery was significant enough for eventual recognition by the Swedish Academy of Sciences and in 1945, twenty years after his discovery, Wolfgang Pauli was awarded the Nobel Prize in Physics "for the discovery of the Exclusion Principle, also called the Pauli Principle."[5]

As a doctoral student at the University of Munich, Wolfgang Pauli was introduced to the current thinking of the structure of the atom by his advisor Arnold Sommerfeld. Danish theoretical physicist Niels Bohr had proposed a model of the atom that placed the electrons in well defined—quantized— orbits around the positively charged nucleus.[6] Each orbital could accommodate up to a maximum possible number of electrons. This maximum followed a simple sequence of 2, 8, 18, 32, or, put another way $2n^2$ where n is a whole number called the *principal quantum number*, which determines the orbital energy. Because the orbitals have different energies, transitions between them involve either a gain or loss in energy. For instance, an electron going from an orbital of higher energy to one of lower energy will emit radiation with an energy equal to the difference between the two orbitals. As described in Chapter 2 this radiation, which may be in the visible part of the electromagnetic spectrum, can be used as a precise measurement of length. Pauli described during his Nobel Lecture, which was delivered on December 13, 1946, that Bohr's quantum atomic model was "somewhat strange from the point of view of classical physics."[7]

Following the publication of Bohr's model in July 1913 much of the subsequent research in this area was directed towards attempts to either correlate the model with classical properties of the electron or to modify it to fit with emerging experimental observations. One result that Bohr's model could not explain was that an atom placed in a magnetic field has a more complicated emission spectrum than the same atom in the absence of a magnetic field. This is the so-called Zeeman effect. Of particular interest—where existing theories failed completely—were the properties of atoms that had many electrons for instance the alkali metals such as sodium and potassium and rare earth elements including dysprosium and erbium. All the elements other than hydrogen have more than a single electron.

As mentioned earlier in Chapter 2, the simple orbit approach with its shortcomings was abandoned in favor of a powerful new theory of atomic structure called quantum mechanics. A key component of quantum mechanics was the motion of electrons.

In the Fall of 1924 while a lecturer at the University of Hamburg Wolfgang Pauli proposed a new "quantum theoretic property of the electron, which I called a 'two-valuedness' not describable classically."[8] This property was the basis for a fourth quantum number that with the previously introduced three quantum numbers *uniquely* define each electron in an atom.[9] In other words, every electron has a unique set of four quantum numbers, not shared with any of the other electrons in that atom. Together the four quantum numbers describe the order in which electron orbitals in an atom are filled providing a valid explanation of the previously unexplainable spectroscopic results. Addition of the fourth quantum number also helped explain why certain materials, for instance those containing iron, are permanently magnetized.

What was difficult for many physicists at the time to accept about the Pauli principle was that there was no physical meaning or property attached to the fourth quantum number. The other three numbers related to the energy, shape, and orientation of the orbital in which an electron moved. It was two young Dutchmen George Uhlenbeck and Samuel Abraham Goudsmit who filled the gap in September 1925 by applying the idea of electron spin.[10] They imagined each electron spinning like a tiny top with direction, angle, and momentum. The fourth quantum number became known as the spin quantum number assigned quantized values of $+\frac{1}{2}$ or $-\frac{1}{2}$.

Niels Bohr, who was now a distinguished Nobel laureate in Physics, added a very supportive footnote to Uhlenbeck and Goudsmit's 1925 *Nature* paper saying that the " ... hypothesis of the spinning electron ... promises to be a very welcome supplement to our ideas of atomic structure." An electron moving in a particular orbital has properties that can be explained by imagining that it is a magnet having a north and a south pole. Just as the direction of current flow around an iron bar determines the direction of the polarity of the magnet induced in the bar, so the direction of the spin of an electron determines its spin quantum number.

In 1927, a year before he was to move from the University of Hamburg to ETH Zürich as Professor of Theoretical Physics, Wolfgang Pauli published a paper describing a magnetic property present in many materials called paramagnetism. Paramagnetic behavior is a weak interaction between a material and an applied magnetic field. Explaining this phenomenon is only possible by recognizing that in some materials each atom can behave as if it were a tiny magnet because of the presence of the spinning electron. If an atom has unpaired electrons, for instance one with spin quantum number + ½ that is not paired with one spinning in the opposite direction the atom will have a net magnetic moment. This moment can align itself with the direction of the applied field. Most chemical elements including the metals aluminum, platinum, and tin as well as many compounds are paramagnetic.

Paramagnetism is not only a weak magnetic effect it is also temporary, disappearing when the applied field is removed. Because of these two factors paramagnetism has no commercial significance but it is an important illustration of how quantum mechanics can be used to explain many of the physical properties of materials. Electron spin is also key to understanding the more commercially relevant magnetic properties of ferromagnetism and ferrimagnetism. Ferromagnets and ferrimagnets are extremely useful materials because they remain permanently magnetized even in the absence of an applied magnetic field.

In ferromagnetic materials for instance the metals iron, cobalt, and nickel, the magnetic dipoles on each atom align through a quantum mechanical exchange reaction where the orientation of one magnetic dipole moment directly influences the spin states of neighboring electrons. This exchange interaction keeps the magnetic moments aligned until a critical temperature, called the Curie temperature is reached.[11] At the Curie temperature the thermal energy seeking to randomize the dipole orientation is stronger than the exchange reaction energy keeping the dipoles aligned.

Although iron is permanently magnetized it is not a particularly useful magnetic material in its elemental form. It is much more powerful and useful when alloyed with boron and the rare earth element neodymium to form the compound neodymium iron boride, $Nd_2Fe_{14}B$, which is a very strong permanent magnet. Neodymium magnets are the strongest commercially available magnets in the world with strengths of 13,000 gauss. For comparison, at its surface the Earth's magnetic field is only ½ gauss, a typical refrigerator magnet is about 100 gauss, and an iron magnet is 9000 gauss. Large neodymium magnets are key components in the enormous wind turbines, which convert mechanical energy into electricity. Using powerful permanent magnets offers many advantages over electromagnets by not requiring a gearbox, which can weigh as much as 80 tons, and not needing a source of electrical power to create a magnetic field. These savings result is a reduction in weight, improved reliability, lower maintenance costs, and improved efficiency. All of which can lower the cost of the electricity produced. The flip side is that permanent magnet generators require large amounts of neodymium and, to a lesser extent, the rare earth elements dysprosium and praseodymium. The permanent magnet content of a turbine generator can be as high as 650 kg per megawatt.[12] A typical wind turbine generates about 2 megawatts and could contain as much as 1300 kg of rare earth permanent magnets. This is just one example of the importance of rare earth elements.

In 1948 French physicist Louis Néel described another type of permanent magnetic behavior, ferrimagnetism.[13] Néel's findings became an important factor in the subsequent development of computer memory and he shared the 1970 Nobel Prize in Physics "for fundamental work and discoveries concerning antiferromagnetism and ferrimagnetism which have led to important applications in solid state physics."[14] Ferrimagnetism, like ferromagnetism, also refers to materials that have a permanent magnetic moment because of unpaired electron spins. Whereas ferromagnets are metal alloys often containing either iron, cobalt or nickel, ferrimagnets, or ferrites, are transition metal oxides containing iron. An example of a ferrite is nickel ferrite, which has the chemical formula $NiO.Fe_2O_3$. The exchange interaction in ferrites is more complex than it is in ferromagnets—it's a *super*exchange mechanism—because it involves the presence of oxygen ions, which separate the iron ions. The outcome of the superexchange is that the magnetic moments of the iron ions, which point in opposite directions exactly cancel each other out. In the example of nickel ferrite the magnetic moment is then due to the presence of unpaired electrons in the nickel ions. Ferrimagnets are not as powerful as many ferromagnets. For instance, a typical ferrite magnet has a strength of about 4000 gauss—less than 1/3 of a neodymium magnet. But it is sufficient enough to permanently store our credit card information on the stripe on the back of the card! That material is barium hexaferrite, $BaO.(Fe_2O_3)_6$.[15]

So, the magnetic properties of materials are determined by the spin of unpaired electrons. For ferromagnets and ferrimagnets the interaction between the spins on the adjacent atoms creates a permanent magnetic moment. For paramagnetic materials the alignment of the magnetic dipole moments only occurs when the material is placed in an external magnet field. Electron spin not only tells us something about the magnetic properties of a material it can also provide very valuable information about the structure of solids. In particular, examining the electron spin state gives us information about defects in solids, which can influence many of the properties of a material.

Scientists can study materials that have unpaired electrons using an analytical tool called electron *paramagnetic* resonance (EPR) or electron *spin* resonance (ESR) spectroscopy. We'll use the term EPR and illustrate how the technique can be used to provide information about structural defects in a solid and its relationship to quantum computing.

Soviet physicist Yevgeny Zavoisky was the first to observe the phenomenon of electron paramagnetic resonance, which he reported in his PhD thesis at Kazan State University in 1944 and then published the following year.[16] Working independently and unaware of the earlier work in Russia, Oxford University physicist Brebis Bleaney developed the phenomenon into a technique that could be used to probe samples containing atoms with one or more unpaired electrons. Bleaney used EPR to study the structures of a wide range of compounds containing transition metals and rare earth elements. In one study published in 1963 Bleaney was able to use EPR results to explain the magnetic properties of the rare earth metal praseodymium based on knowledge of its crystal structure.[17] Praseodymium, as noted earlier, is present in small amounts in the permanent magnets used in wind turbines.

Much like Brebis Bleaney's work at Oxford, research in electron paramagnetic resonance was primarily confined to university laboratories using home-built apparatus. That was until the late 1950s when commercial instruments such as the JES-1 was released by Japanese scientific instrument maker JEOL, Ltd. The company already had considerable experience building advanced scientific instrumentation having produced its first transmission electron microscope in 1950 and its first nuclear magnetic resonance (NMR) spectrometer in 1956.

An electron paramagnetic resonance spectrometer shares many similarities with NMR. The sample to be analyzed is placed inside a cavity where it is exposed to microwave radiation typically held at a frequency of 9.5 gigahertz. This is a much higher frequency than in a household microwave oven, which operates at a frequency of 2.45 gigahertz. The microwave cavity is located in the middle of an electromagnet, whose field can be varied until the microwave energy is absorbed by the unpaired electron causing it to move to a higher energy state. We say that the magnetic field "tunes" the two spin states until their energy difference is equal to the applied microwave radiation. The output, the absorption spectrum, is a plot of absorption versus magnetic field. Although EPR cannot be used to determine the composition of a material or the presence of certain elements it can provide important information about the environment of a particular atom in a material including the presence of structural defects.

To understand how EPR can be applied in materials science, we can look at the example of defects in amorphous silicon. As mentioned in Chapter 7, silicon is the most widely used material to produce solar cells. More silicon is used in the photovoltaic industry than in the manufacture of computer chips. Depending on the demands of the application and considerations of cost the cells can be fabricated from either single crystal silicon, or poly-crystalline material (called polysilicon), or an amorphous or glassy form of silicon. If amorphous silicon is used it is often in the form of a thin film in low-efficiency, low-cost, low-power applications. Amorphous silicon cells are considerably cheaper than their crystalline counterparts. The medium price for an amorphous silicon panel is $0.84 per watt, while for a single crystal it is $1.1 per watt.[18]

Electron paramagnetic resonance spectroscopy has been used to investigate how paramagnetic centers—defects—formed during cell degradation influence the overall device efficiency. Figure 8.2 illustrates a possible defect in amorphous silicon corresponding to a strong absorption at a magnetic field between 3510 and 3520 gauss. The incoming light has broken a silicon–silicon bond creating a silicon vacancy (an unoccupied silicon lattice site) leaving behind what is called a "dangling bond". These defects have been shown to reduce the efficiency of amorphous silicon solar cells because they can capture electrons thereby preventing them from contributing to the overall electrical current.

Successful development of quantum computers will rely on our ability to control and manipulate electron spin. Since the 1990s, scientists have been able to gain increasingly better control over electron spin. Not only how to use it diagnostically, as in EPR spectroscopy, but also how to manipulate it. A particularly striking study showing spin manipulation at the atomic level combined a scanning tunneling microscope, an STM, with low-temperature EPR to allow researchers to create single-atom magnets.[19]

Figure 8.3 illustrates the experimental set-up, which shows single holmium and iron atoms on a magnesium oxide surface. The magnesium oxide film is grown onto a silver substrate to a thickness of about 0.4 nm. Using a magnetized iridium tip the magnetic state of the holmium atom can be

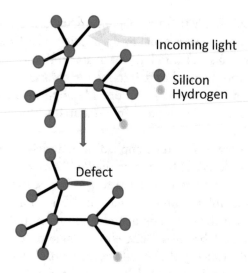

Incoming light

Silicon
Hydrogen

Defect

Fig. 8.2 Diagram illustrating a possible defect structure in amorphous silicon determined by electron paramagnetic resonance spectroscopy. The defect was induced by the incident radiation

controlled—written. The magnetic state of the holmium atom can also be read by the scanning tunneling microscope. So, it is possible to *read* and *write* magnetically stored information by manipulating electron spin. One important consideration is that the sample must be kept very cold. It is held at a temperature below −263 °C where the holmium is ferromagnetic.[20] (As mentioned earlier ferromagnetic materials remain magnetized even in the absence of an applied field.) The research team led by the IBM Almaden Research Center in San Jose California found that the holmium atom retained its magnetic information over many hours. Using electron paramagnetic resonance on an adjacent iron atom, acting as an atomic-sized sensor, it was possible for the researchers to prove the magnetic origin of the switching. It is the direction of electron spin of the holmium atom that is being switched.

There are several different approaches to fabricating the qubits, which are at the heart of a quantum computer. Most of the existing quantum computers use superconducting qubits. When cooled below a critical temperature, superconductors can carry an electrical current with no loss—zero resistance. This unusual property is because at low temperatures, often very close to the absolute zero of temperature, the electrons in the superconducting material bind together forming Cooper pairs. Because of entanglement effects the quantum states in superconductors are longer lived than in conventional materials. A further peculiarity of the quantum world, which we make use

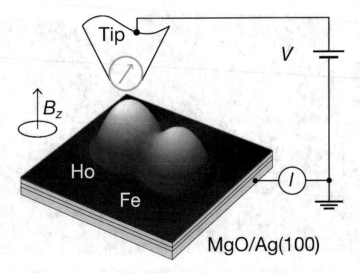

Fig. 8.3 Topographic image of a holmium (Ho) and an iron (Fe) atom on a layer of magnesium oxide (MgO) on silver (Ag). The magnetic states of Ho are controlled and probed with a scanning tunneling microscope tip, which is fabricated of iridium. B_z indicates the direction of the magnetic field (Reprinted from Natterer FD, Yang K, William P, Willke P, Choi T, Greber T,Heinrich AJ, Lutz CP (2017) Reading and writing single-atom magnets. Nature 543:226–228 with permission from Springer Nature)

of in the fabrication of qubits is that Cooper pairs can tunnel through very thin insulating regions. When two pieces of superconductor are separated by a thin insulating layer, for instance two metal layers separated by a thin oxide film the device is called a Josephson junction named after University of Cambridge physicist Brian David Josephson. The 1973 Nobel Prize in Physics was awarded to Josephson "for his theoretical predications of the properties of a super current through a tunnel barrier, in particular those phenomena which are generally known as the Josephson effects."[21]

In 1962, while still a graduate student at Cambridge Josephson predicted that electrons in superconductors could pass through insulating barriers in a way that would be disallowed according to classical physics. The first such junction made by Brian Josephson consisted of a niobium wire surrounded by a blob of solder separated by a thin oxide layer. Modern day Josephson junctions of the type used in quantum computers are much more controlled structures than the early devices made at Cambridge. An example of a superconducting qubit is shown in Fig. 8.4. It consists of two layers of aluminum separated by an insulating film of aluminum oxide. The two Josephson junctions are indicated by arrows and sometimes referred to as a Cooper-pair box.

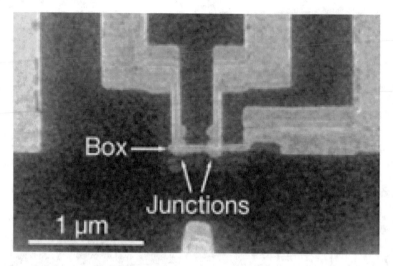

Fig. 8.4 Electron microscope image of a superconducting qubit. The circuits are made of aluminum films. The Josephson junction consists of insulating aluminum oxide tunnel barriers between two electrically conducting layers of aluminum (Reprinted from Ladd TD, Jelezko F, Laflamme R, Nakamura Y, Monroe C,O'Brien JL (2010) Quantum computers. Nature 464:45–53 with permission from Springer Nature)

An alternative approach to forming superconducting qubits is to use atomic sized defects in a material. All materials contain vacancies, locations in the structure where an atom is missing. Some materials also contain misplaced or dopant atoms where an atom is sitting on the "wrong" site. In certain quantum materials these defects trap electrons allowing researchers to access and control their spins. Unlike superconductors, these dopant qubits don't always need to be at ultra-low temperatures, some can even operate at room temperature. An example of a material that is being studied as a quantum material is diamond. Diamonds have extremely small numbers of vacant sites in their structure because the bonding between carbon atoms is incredibly strong, but it is possible to substitute nitrogen atoms for carbon atoms creating a "nitrogen-vacancy" center. These defects are also referred to as color centers because they can colorize crystals. For instance, nitrogen-doped diamonds have a grayish-brown coloration. Undoped "pure" diamonds are colorless.

The nitrogen-vacancy center consisting of a substitutional nitrogen at a lattice site neighboring a missing carbon atom is illustrated in Fig. 8.5. The spin state of the defect can be manipulated using an applied microwave field. Other materials that have structures similar to that of diamond are also in the picture for dopant qubits. Possible alternatives to diamond include the less expensive synthetic ceramic materials aluminum nitride and silicon carbide.

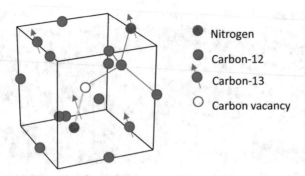

Fig. 8.5 The atomic structure of a nitrogen-vacancy (N-V) center in the diamond lattice. The side edge of the cube is 0.36 nm

Malcolm Carroll and Thaddeus Ladd have made a case for building qubits out of silicon.[22] One of the reasons for using silicon is the considerable experience of working with this material in conventional microelectronics technology. As we saw in Chapter 4 there are established methods for growing ultra-high purity single crystals and fabrication techniques that have allowed integration of billions of nanoscale features onto a single chip. Although quantum computers and classical computers are fundamentally different, being able to leverage decades of work could give silicon an advantage in quantum technology. Keiji Ono of the Advanced Device laboratory of RIKEN in Japan explains: … we are attempting to develop a quantum computer based on the silicon manufacturing techniques currently used to make computers and smart phones. The advantage of this approach is that it can leverage existing industrial knowledge and technology."[23]

Figure 8.6 shows a MOS silicon qubit fabricated at Sandia National Laboratories in Albuquerque, New Mexico. The device structure shares some similarities with the silicon MOSFET illustrated earlier in Fig. 8.1. The qubit stack consists of a single crystal n-type silicon substrate, a 35 nm thick silicon dioxide, silica, gate oxide topped with a 200 nm thick polycrystalline silicon gate.

While superconducting qubits are currently leading the way, it is not settled on which technology will prove to be the best and eventually prevail. Despite the success in identifying and fabricating different forms of qubit, there are considerable challenges to quantum computing remaining that will likely continue well into the middle of this century and possibly beyond.

To illustrate the complexity of working in the quantum regime we can compare the operation of a conventional computer with a quantum computer. An essential action of our laptops and desktop computers is the ability to create and store information in the form of bits. This information

MGW   GW-SWAG   MSWAG

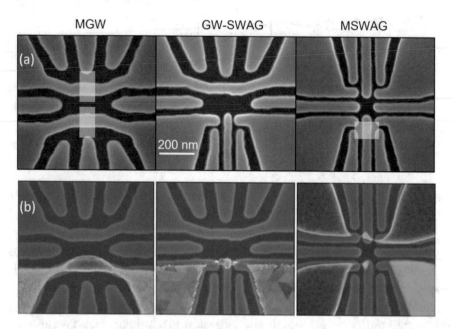

Fig. 8.6 **a** Scanning electron microscope (SEM) images of three device designs. The light gray are the poly-silicon gates, and the yellow regions are where donors are implanted. **b** Simulated electron density (red regions) for each of the devices using operational voltages. The terms MGW, GW-SWAG, and MSWAG refer to the different designs (Reprinted with permission from Rudolph M, Sarabi B, Murray R, Carroll MS,Zimmerman NM (2019) Long-term drift of Si-MOS quantum dots with intentional donor implants. Scientific Reports 9:7656 under the terms of the Creative Commons CC BY license)

should remain stored indefinitely allowing us to access it whenever we want until such time that we deliberately change it or delete it. If we were not able to do that simple operation our computers would have very limited use functioning more as an expensive electronic typewriter. In quantum computers, qubits are very sensitive to their environment and can be unstable causing the stored information to change on its own. This property is known as coherence. Right now quantum computers have low coherence, meaning that errors are high. Danna Freedman a professor of chemistry at Northwestern University has identified the basic problem: "Making sure that you get the right answer all of the time is one of the biggest hurdles in quantum computing."[24] For a user this should be an absolute minimum requirement of any computer.

So, right now there are multiple challenges with quantum computers. The low coherence time and the propensity to produce errors are among the

Fig. 8.7 Google's Sycamore chip is composed of 54 qubits, each made of superconducting loops

biggest challenges that scientists and engineers need to address. For the materials scientist there are two questions: What are the current limitations of the materials used to make quantum computers and how can we improve on them?

In 2019 scientists at Google led by University of California, Santa Barbara physicist John Martinis reported in a *Nature* paper that they had achieved quantum supremacy. A quantum computer had been able to perform a specific calculation that is beyond the capabilities of a state of the art, classical, computer.[25] It was estimated that the same calculation would take even the best conventional computer 10,000 years to complete. The Google device, which consisted of 54 superconducting Josephson junction qubits is shown in the photograph in Fig. 8.7 took only 3 min and 20 s. (Only 53 of the qubits were used as one of the 54 was found to be broken.) While this result is exciting and marks a milestone on the path to full-scale quantum computing a general-purpose quantum computer would need not 54 qubits, but maybe as many as one million. The more qubits that are linked, the harder it is to maintain their fragile states while operating. And this presents a significant technical challenge.

David Awschalom, a quantum physicist at the University of Chicago has compared where we are today with quantum computers to where we were in the 1950s beginning our exploration of transistors.[23] We have come a very long way with transistor technology from the first ever device pictured in Fig. 8.8 made at Bell Labs in 1947. We persevered with transistors because their potential was obvious. We will persevere with quantum computing because they have the potential to solve problems that

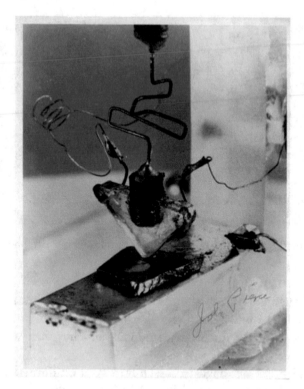

Fig. 8.8 Bell Laboratories' first transistor with a germanium crystal. Courtesy of the Computer History Museum (Image courtesy of the Computer History Museum)

conventional computers, even the most powerful supercomputers, can't. The discovery of new medicines and advanced materials, not by conducting a series of trials or time-consuming experiments but by simulating chemical and physiological processes within the computer, are among the possible, and exciting, benefits of quantum computers.

Notes

1. Lilienfeld JE (1930) Method and apparatus for controlling electric current, US 1745175. Patent filed in United States October 1926, issued January 1930.
2. The metal–oxide–semiconductor field-effect transistor (MOSFET) is a three terminal device comprising a source, gate, and drain. Depending on the voltage applied to the gate electrode, current can be made to flow or not between the source and the drain. The MOSFET was invented in 1959 by Bell Lab's Mohammed M. Atalla and Dawon Kahng and is the basic

building block of modern electronics. The device is covered by two U.S. patents 3,206,670 filed by Atalla and 3,102,230 filed by Kahng.

3. Ladd TD, Jelezko F, Laflamme R, Nakamura Y, Monroe C, O'Brien JL (2010) Quantum computers. Nature 464:45–53.

4. Shor PW (1994) Algorithms for quantum computation: discrete logarithms and factoring. In: Proceedings 35th annual symposium on foundations of Computer Science, pp 124–134. Since 2003 Shor has been a Professor of Applied Mathematics at the Massachusetts Institute of Technology.

5. The Nobel Prize in Physics 1945. NobelPrize.org. Nobel Media AB 2021. Fri. 28 May 2021. https://www.nobelprize.org/prizes/physics/1945/summary/.

6. Bohr N (1913) I. On the constitution of atoms and molecules. Lond, Edinb, Dublin Philosophical Magazine and Journal of Science 26:1–25. The first in a series of three papers introducing a model of the atom that became known as the Bohr atom. Bohr was awarded the 1922 Nobel prize in Physics for his work in this area.

7. Wolfgang Pauli Nobel Lecture, December 13 1946. Available at https://www.nobelprize.org/uploads/2018/06/pauli-lecture.pdf.

8. Pauli W (1925) Uber den einfluss der geschwindigkeitsabhängigkeit der elekronmasse auf den Zeemaneffekt. Zeitschrift für Physik 31:373–385. Translation of the title is: About the influence of the speed dependence of the electron mass on the Zeeman effect. A subsequent paper describes the formulation of the exclusion principle, which is almost always referred to now as the Pauli principle: Pauli W (1925) Uber den zussammenhang des abschlusses der elektrongruppen im atom mit der komplexstruktur der spektren. Zeitschrift für Physik 31:765–783. Translation of the title is: About the connection between the completion of the electron groups in an atom with the complex structure of spectra.

9. There are four quantum numbers used to identify an electron in an atom. The principal quantum number, n, determines the effective volume of an electron orbital (analogous to the shell in the Bohr model); the orbital quantum number, l, determines the shape of the orbital; the orbital-orientation quantum number, m_l, determines the orientation of the orbital with reference to an external magnetic field; m_s is the spin quantum number. Arnold Sommerfeld introduced the second and third quantum numbers.

10. Uhlenbeck GE, Goudsmit S (1925) Ersetzung der hypothese vom unmechanischen zwang durch eine forderung bezüglich des inneren verhaltens jedes einzelnen elekrons. Die Naturwissenschaftem 13:953–954; Uhlenbeck GE, Goudsmit S (1926) Spinning electron and the structure of spectra. Nature 117:264–265. Neither Uhlenbeck nor Goudsmit had doctoral degrees. Uhlenbeck, 24, had a masters ('doctorandus') and Goudsmit, 23, was a graduate student. They were students of Professor Paul Ehrenfest.

11. The Curie temperature is named after Pierre Curie who demonstrated that materials lose their magnetic properties above a certain temperature. For instance, the T_c for iron is 770°C, but over 1100°C for cobalt. For the

compound europium oxide, EuO, $T_c = -204°C$. The Curie temperature is not to be confused with the critical temperature for superconductivity, which also is represented by the symbol T_c.

12. https://www.windpowermonthly.com/article/1519221/rethinking-use-rare-earth-elements

13. Néel L (1948) Propriétés magnétiques des ferrites; ferrimagnétisme et anti-ferromagnétisme. Annals of Physics 3:137–198.

14. Louis Néel – Facts. NobelPrize.org. Nobel Media AB 2021. Tue. 15 Jun 2021. https://www.nobelprize.org/prizes/physics/1970/neel/facts/.

15. Barium hexaferrite is used for credit card data storage because it is very stable with excellent corrosion resistance making it more reliable than metal magnets.

16. Zavoisky E. (1944) Paramagnetic absorption in perpendicular and parallel fields for salts, solutions and metals. PhD Thesis. Kazan State University. Zavoisky, E (1945) Spin-magnetic resonance in paramagnetics. Fiz Zhurnal 9:211–245.

17. Bleaney B (1963) The magnetic properties of praseodymium metal. Proceedings of the Royal Society of London A 276:39–50.

18. Cotfas DT, Cotfas, PA (2019) Comparative study of two commercial photovoltaic panels under natural sunlight conditions. International Journal of Photoenergy. Article ID 8365175.

19. Natterer FD, Yang K, Paul W, Willke P et al (2017) Reading and writing single-atom magnets. Nature 543:226–228.

20. Rhodes BL, Legvold S, Spedding FH (1958) Magnetic properties of holmium and thulium metals. Physical Review 109:1547. Like many materials the magnetic properties of holmium are a function of temperature. It is paramagnetic above 133°K, antiferromagnetic between 20°K and 133°K, and ferromagnetic below 20°K. It is necessary that to store information magnetically the material must be in its ferromagnetic state.

21. Brian D. Josephson – Facts. NobelPrize.org. Nobel Prize Outreach AB 2021. Fri. 2 Jul 2021. https://www.nobelprize.org/prizes/physics/1973/josephson/facts/

22. Ladd, TD, Carroll MS (2018) Silicon qubits. Encyclopedia of Modern Optics 1:467–477.

23. Ono K, Giavaras G, Tanamoto T, Ohguro T, Hu X, Nori F (2017) Hole spin resonance and spin–orbit coupling in a silicon metal–oxide–semiconductor field-effect transistor. Physical Review Letters 119:156802.

24. Quoted in: https://www.energy.gov/science/articles/creating-heart-quantum-computer-developing-qubits.

25. Arute F, Arya K, Babbush R, Bacon D et al (2019) Quantum supremacy using a programmable superconducting processor. Nature 574:505–510.

9

Promises Unmet

In 1986 Johannes Georg Bednorz and Karl Alexander Müller of IBM's Zürich Research Laboratories in Rüschlikon, Switzerland made a discovery that was significant enough to earn them both an equal share of the 1987 Nobel Prize in Physics for "their important break-through in the discovery of super-conductivity in ceramic materials."[1] Specifically, what the two scientists had found was that a compound containing the elements barium, lanthanum, copper, and oxygen (Ba-La-Cu–O) that had been synthesized two years earlier by French chemists Claude Michel and Bernard Raveau at the Université de Caen became superconducting—lost all its electrical resistance—when cooled to the unexpectedly warm temperature of 30 K—thirty degrees above abso-lute zero.[2] A temperature of 0 K is the lowest that it is theoretically possible to achieve. Michel and Raveau had measured many of the properties of their material and even determined it was a very good metallic-like conductor, but they had not attempted to measure whether it would superconduct.[3]

The temperature at which a material passes into the superconducting state—transforming from a normal conductor or even an insulator into a superconductor—is labelled the critical temperature, T_c. A critical temper-ature of 30 K was so surprising to Bednorz and Müller that they rather cautiously titled their landmark paper, "Possible high T_c superconductivity in the Ba-La-Cu–O system" rather than a bolder and more concise "High T_c superconductivity in the Ba-La-Cu–O system." The reason for their caution was, in part, that the previous record value for critical temperature was only 23 K obtained by the niobium germanium alloy Nb_3Ge. For an

© The Author(s), under exclusive license to Springer Nature
Switzerland AG 2023
M. G. Norton, *A Modern History of Materials*,
https://doi.org/10.1007/978-3-031-23990-8_9

oxide, the highest critical temperature that had been previously measured was just 13 K—less than half that of the compound Bednorz and Müller were studying.

The results coming out of Switzerland also indicated that the accepted theory used to explain superconductivity, the BCS theory due to the Nobel Prize winning trio of John Bardeen, Leon Cooper and John Schrieffer may not apply to ceramics and that critical temperatures beyond the limits set by BCS theory might one day be possible.[4] According to the three University of Illinois physicists, negatively charged electrons could under certain conditions at very low temperatures overcome their mutual repulsion to pair up forming so-called Cooper pairs, named as mentioned in Chapter 8 after Leon Cooper. It is these electron *pairs* that are responsible for carrying a supercurrent. They are not subject to the resistance experienced by single electrons. So, what Bednorz and Müller had demonstrated was the world's first ever high-temperature superconductor, a material that might involve Cooper pairs but did not necessarily follow the BCS rules.[5]

Superconductivity is one of the most remarkable of all the physical properties that can be exhibited by a material. When cooled below its critical temperature a superconductor loses all resistance to the flow of electricity. An induced current will flow indefinitely without inhibition or decay. This behavior is very different to what we see above the critical temperature where the resistance to the flow of electrons causes a material to heat up. In some cases, this resistance can be so large that the material will glow with a color changing from a dull red to a bright painful white as the temperature increases.[6]

At the same time that all electrical resistance is lost, a superconductor demonstrates what is known as the Meissner effect where a magnetic field cannot—or at best only partially—penetrate the surface of the material.[7] As a superconductor is cooled through its transition temperature a magnet placed on the material's surface will spontaneously levitate. So, superconductivity is not only an electrical phenomenon it is also a magnetic property. And both behaviors immediately suggested exciting and ground-breaking possibilities for superconductors that might not be restricted to the frigid limits imposed by BCS theory.

One might imagine the power lines that crisscross our nation carrying electrical power over many hundreds of miles with no loss at all because they are made of a superconducting wire rather than the typical metal alloy conductors such as aluminum-clad Invar. With traditional wires it is not unusual for 20% of the power generated to be wasted due to the inherent resistance of the metal. But for a superconductor that loss would be zero, leading to cheaper and greener electricity. Or imagine instead a futuristic streamlined train levitated above superconducting rails travelling at previously unobtainable speeds made possible because of the complete absence of friction.

In addition to these grand infrastructure projects there were numerous other proposed applications for "high T_c" superconductors including ultrafast electronic circuits that might one day outperform and replace the ubiquitous silicon chip.

We are now more than three decades past Bednorz and Müller's original publication on high-temperature superconductivity. Their findings moved very rapidly from the "*possible*" to the *definite* as the results were widely confirmed and reproduced by scientists around the world. Unfortunately, none of the grand applications that would have undoubtedly been game changers have yet been realized. Public interest and that of agencies that might fund the necessary research into superconductor technology has waned. There are several reasons why high-temperature superconductors have so far been unable to match their much-hyped promise:

- 30 K, though a high temperature in the world of superconductivity, is still extremely cold. To reach these temperatures requires the use of liquid helium, which boils at 4 K (-269°C).
- Ceramic superconductors, much like the vast majority of ceramics, are very brittle, particularly when cold.
- It is very difficult to draw ceramics into long thin wires, which would be the form most useful for carrying electrical power.

In addition to speculating on the technologies that might be realized with high-temperature superconductors, the scientific community invested considerable resources to overcome the challenge of the very low critical temperatures that were required to enable the superconducting state. Over 100,000 scientific papers on high-temperature superconductivity were published in the two decades since Bednorz and Müller's publication. Another 100,000 were added to the scientific literature in the next five years. Millions of dollars were invested by almost every scientific research agency in the world. In the United States, federal government support for research in high-temperature superconductors went from $45 million in 1987 (when Bednorz and Müller were awarded the Nobel Prize) to $130 million just three years later (over $260 million in today's dollars).[8]

Much of this research was focused on the goal of moving the temperature at which a material becomes superconducting from Bednorz and Muller's

30 K to the much warmer and technically viable 77 K. At this temperature the most abundant gas on earth, nitrogen, is a liquid.

The approach that was taken by almost every group funded to work in this area was to look at the palette of elements, the Periodic Table of Elements, take the Ba-La-Cu–O compound as a template and begin substituting one element for another. Without a firm theoretical understanding of how these new ceramic superconductors worked the only way to approach increasing the critical temperature was to adopt the so-called Edisonian method of trial and error.[9] Such an approach was confirmed at a superconductivity workshop in 1988 held in Copper Mountain, Colorado: "At the extreme forefront of research in superconductivity is the empirical search for new materials."[10]

Although there are 118 known elements, the number of possible substitutions into the Ba-La-Cu–O template is limited. For example, we cannot substitute in any of the Noble Gases because they do not readily form compounds with other elements except in certain cases with fluorine. That restriction removes six elements from consideration. Also, we cannot substitute any of the elements with atomic numbers larger than that of uranium—92—because they are radioactive, exist for an incredibly short time, don't exist in nature, or all three. This restriction eliminates another 27 elements. Eliminating all the other radioactive elements such as francium, radium, and polonium further reduces the options. But the remaining 78 elements were, in 1986, more or less fair play. The choices quickly became further limited, however, when the critical—and irreplaceable—role of copper and oxygen were identified. So, with balances, spatulas, pestles-and-mortars, furnaces, and elaborate testing equipment thousands of the world's scientists attempted to get a high-temperature superconductor that would work at the temperature of liquid nitrogen. Or above.

The group that succeeded first was led by Paul Ching-Wu Chu at the University of Houston and Maw-Kuen Wu at the University of Alabama in Huntsville.[11] Their unique combination of elements was yttrium, barium, copper, and oxygen. Three of the same elements used by Bednorz and Müller, but with one rare-earth element yttrium replacing another rare-earth element lanthanum. When the ratio of the elements was adjusted to give the chemical formula $Y_1Ba_2Cu_3O_7$—as simple as "123"—the material would superconduct at the comparatively balmy 93 K, well above the 77 K temperature of liquid nitrogen. This discovery by Chu and Wu further fueled the frenzy of investment and research activity in superconductivity because liquid nitrogen is inexpensive. A liter costs about $0.45, in contrast to a liter of liquid helium at around $15.00, and liquid nitrogen is easy to handle. The door truly seemed open to the creation of new devices and applications based on superconductivity that had the potential to change the world. With this breakthrough scientists talked earnestly about the transmission of electrical power over enormous distances without any loss, tiny supercomputers, and the "flagship" application, high-speed levitating trains that would usher in a new era of travel. In many laboratories around the world scientists whispered about whether it was possible to get a superconductor that would operate at room temperature, a temperature typically defined as between 288 and 300 K.

Substitution after substitution was tried, add some calcium or strontium for barium; use bismuth, mercury, or thallium instead of yttrium; change the ratio of calcium to strontium; add even more thallium. Slowly the critical temperature inched higher eventually reaching a record value for a doped copper oxide or cuprate superconductor, which stands at 134 K with the composition $HgBa_2Ca_2Cu_3O_8$.[12] Thirty-five-plus years after publication of the paper that launched the field of high-temperature superconductivity and almost thirty years since Paul Chu and colleagues at the University of Houston described superconductivity in the mercury containing compounds, we have been unable to shift the critical temperature of cuprate superconductors any closer to room temperature.

But room temperature superconductivity has been reported in other materials systems that are not part of the cuprate family. In other words, compounds that do not require copper and oxygen.

All the superconductors discovered so far with critical temperatures above 200 K have been hydrogen-rich materials exposed to extremely high pressures. In a *Nature* article in 2019 an international team led by the Max Planck Institut für Chemie in Mainz, Germany found superconductivity in the compound lanthanum hydride, LaH_{10}, at a temperature of 250 K when the material is compressed to 170 gigapascals.[13] This is a very different system to the oxide superconductors and to achieve the high value of critical temperature requires the application of a pressure equivalent to that in the Earth's Mantle around a thousand kilometers below the surface.[14] More recently a group led by the University of Rochester in New York measured room-temperature superconductivity in a compound containing the three elements carbon, sulfur, and hydrogen.[15] The highest critical temperature they obtained was 287.7 K (a pleasant Spring afternoon temperature) when the material was pressurized to 267 gigapascals—a pressure found deep in the Earth's Mantle close to the Core.

These newly discovered superconductors—the hydrides—much like their cuprate counterparts present significant challenges when considering possible commercialization outside of the university laboratory. Not the least of which is the complexity and cost of working at the extremely high pressures necessary for these materials to enter the superconducting state.

Even after the work of Chu and Wu that led to oxide superconductors that worked at liquid nitrogen temperatures there remained significant challenges in realizing commercial uses for high-temperature superconductors. In most of the proposed large-scale applications the material was required to be in the form of long thin wires. Drawing metals, for instance aluminum and copper, into wires is relatively straightforward and can even be performed without the use of heat. The metals readily deform when pulled through a die or a series of progressively smaller dies, which reduce the cross section until the desired wire thickness is formed. It is possible to reach diameters down to fractions of a millimeter (just a few micrometers.) This type of processing is impossible to do with ceramics because they are brittle, especially when cold, and do not readily deform plastically as most metals and metal alloys do. An alternative approach that has been used is to pack powders made from superconducting oxides into hollow silver tubes, which can be pressed into flat tapes. These tapes generally lack the flexibility of metal wires. In addition, the boundaries between the powder particles, the grain boundaries, can act as weak links significantly reducing the amount of current that can flow through the tapes before they lose their superconducting powers.[16] The latter limitation has been overcome using a high-temperature superconductor in the bismuth strontium calcium copper oxide (BSCCO, *bisko* family, which form plate-like grains as shown in Fig. 9.1

In BSCCO superconductors the grains stack in a somewhat ordered way, resembling the structure of a slate wall, such that the weak link behavior of the grain boundaries is significantly reduced. These superconducting tapes are attractive because they can carry much larger currents than conventional cables of the same diameter.

Some modest projects using superconducting tapes containing the BSCCO superconductor cooled by liquid nitrogen as transmission power cables have been demonstrated. One of the earliest studies examining the feasibility of superconducting cables was conducted in Denmark as a direct comparison with an aluminum cable.[17] A cable length of 4,050m (a little over 2½ miles) was modelled. The outcome of the study showed that energy savings using the superconducting cable system could be up to 40% and at a comparable cost to metal cables. Despite these encouraging findings it appears the system was never put into commercial service.

The first system operating at transmission voltages—real-life conditions—was the nearly half mile long superconducting cable forming part of the Long Island Power Authority's grid network in Lake Ronkonkoma, New York.

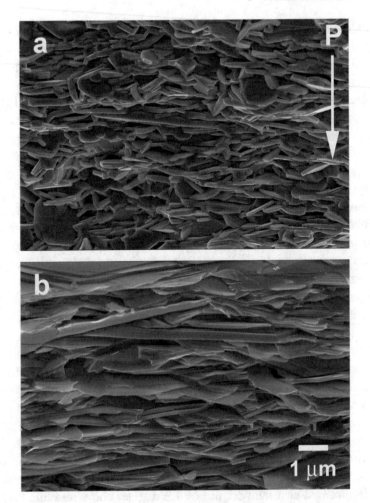

Fig. 9.1 Scanning electron microscope image of a *bisko* superconductor. The plate-like grains and how they stack on top of each other reduces the loss of current that flows through the material. P indicates the direction of the applied pressure used during processing of the superconductor (Reprinted from Kahraman F, Sotelo A, Madre MA, Diez JC, Ozkurt B, Rasekh Sh (2015) Relationship between microstructure and superconducting properties in hot-pressed Bi-2212/Ag ceramic composites. Ceramics International 41:14924–14929 with permission from Elsevier)

Energized in 2008 the cable had the capacity to serve about 300,000 homes. The total project cost was almost $47 million ($57 million in today's dollars). Half the funds came from the Department of Energy, the other half from the private sector.

In 2014 the "AmpaCity" project installed a more than one kilometer long superconducting cable to supply about 10,000 households in the western German city of Essen. This was the longest high-temperature superconductor cable in the world. Despite its complex multilayer construction including the outer cooling jacket for the liquid nitrogen coolant, the superconducting cable can transport five times the electricity as that of a similarly sized copper cable—and with far fewer electrical losses.

Davide Uglietti at the Swiss Plasma Center part of the Ecole Polytechnique Fédérale de Lausanne (EPFL) highlighted the challenges associated with commercialization of high-temperature superconductors despite their ability to operate at the relatively warm temperature of liquid nitrogen. These challenges included their short piece lengths, limited production capacity, and high costs.[18] Piece lengths for coated BSCCO superconductor tapes are in the order of 1 km but can be much less depending on the material and the structure of the tape. As a comparison low-temperature superconductor wires as long as 200 km are routinely produced. The price of superconductor materials varies greatly depending on the material and the quantities required by the customer. A useful comparison is to compare the cost in units of dollars per kiloamp-meters as a function of the magnetic field. The magnetic field of a top end research grade medical imaging system is about 11 T. Using a niobium tin superconductor the price would be less than $15 per kiloamp-meter. A BSCCO tape could easily cost ten times more.

To put Bednorz and Müller's discovery into context and the frenzy of research that it launched it is relevant to go back to the origin of superconductivity and the work in the field of ultra-cold physics, which was attempting to achieve temperatures as close to the absolute zero of temperature as possible.

The history of superconductivity begins in 1911 in Leiden in the Netherlands where Heike Kamerlingh Onnes measured the electrical resistance of mercury, the only liquid metal at room temperature, and found that its resistance disappeared at a temperature of 4.2 degrees above absolute zero. This result, much like that of Bednorz and Müller seventy five years later, was completely unexpected. So much so that Kamerlingh Onnes, who held the Physics Chair at Leyden University, was awarded the 1913 Nobel Prize in Physics "for his investigations on the properties of matter at low temperatures which led, inter alia, to the production of liquid helium."[19]

Kamerlingh Onnes coined the term "superconductivity" to describe the state where a material loses all electrical resistance; an induced current will persist indefinitely. The metallic superconductors, mercury, tin, and lead that were the subject of his research and the later discovered superconducting alloys including niobium zirconium (NbZr) and niobium titanium (NbTi) became known as low-temperature superconductors because they required cooling with liquid helium to operate.

It took almost sixty years for large scale applications to be found for the superconducting metallic alloys. Superconducting magnets for nuclear fusion reactors and particle accelerators were manufactured in the 1970s incorporating wires made from low-temperature superconductors. Around the same time powerful superconducting magnets wound using either niobium titanium or niobium tin wires were required for magnetic resonance imaging (MRI) and, its closely related chemical analysis technique, nuclear magnetic resonance (NMR) spectroscopy.

In the early 1980s there were just twelve MRI instruments in service, now there are approximately 50,000 such machines in hospitals, clinics, and laboratories around the world. Over 35,000 of which use superconducting magnets. MRI has become an essential imaging technology, an important part of a physician's toolbox. Its detailed three-dimensional anatomical images can be used to detect the onset of disease, assist with diagnosis, and monitor disease progression and the effectiveness of treatment. The essential component of all MRI instruments is the superconducting magnet, which produces a strong magnetic field that forces protons (the nuclei of hydrogen atoms) in the body to align with the direction of the applied field. When a radiofrequency (RF) current is pulsed through the patient, the protons, which have been aligned by the magnetic field, are stimulated and spin out of equilibrium, straining against the pull of the magnetic field. When the radio frequency field is turned off the MRI sensors detect the energy released as the protons realign themselves with the magnetic field. The time it takes for this realignment as well as the amount of energy released changes depending on the environment and the chemical nature of the molecules that contain the stimulated protons. Physicians are able to tell the difference between various types of tissues based on their response. The benefits of using superconducting magnets include sharper images, shorter scan times, and higher patient throughput than machines with conventional magnets.[20] Depending on the strength of the magnetic field an individual MRI instrument can cost well in excess of one million dollars.

While MRI using superconducting magnets has revolutionized medical imaging, NMR spectroscopy is an invaluable analytical tool for the chemist. The technique was first demonstrated independently in 1946 by Stanford University physicist Felix Bloch and Harvard University's Edward Mills Purcell, who shared the 1952 Nobel Prize in Physics "for their development of new methods for nuclear magnetic precision measurements and discoveries in connection therewith."[21] Although the first commercial instruments used conventional electromagnets and permanent magnets, by the 1960s super-conducting magnets were the standard leading the way eventually to modern compact benchtop spectrometers.

NMR works in a very similar way to MRI. A powerful and stable magnetic field is applied to the sample, which is most commonly in the form of a liquid, but solids can also be used in some instruments. The nuclei of certain atoms (such as a hydrogen nucleus ^{1}H as in MRI or a carbon nucleus ^{13}C) will behave as tiny magnets aligning themselves with the direction of an external magnetic field. If the sample is now exposed to radio frequency waves the nuclei will resonate at their own specific frequencies. When the atoms return to their original positions, they reemit radio waves with frequencies that are characteristic of the atoms and molecules that contain those resonating nuclei and their environments. The radio waves are detected, measured, and converted into a spectrum; a graph that plots the intensity of the signal received by the detector against a parameter called the chemical shift. Intensity is simply the number of nuclei that resonate at each specific frequency. The chemical shift tells us about the different environments that the nuclei are in—their neighbors and their location. NMR provides critical information that has impacted many fields, none more so than medicine, particularly in the field of cancer research, where it is used among other things for the development of smart delivery systems for cancer-treating drugs.

As more powerful superconducting magnets have become available, scientists can use NMR to study increasingly complex molecules and obtain more detailed three-dimensional information about molecular structure.

In 2021 the most powerful commercial NMR instrument of its kind was installed at the Utrecht Science Park at Utrecht University in the Netherlands. The superconducting magnet is a unique hybrid design combining both high- and low-temperature superconductors. The low-temperature superconductor is a niobium tin alloy ($NbSn_3$) while the high-temperature superconductor is an yttrium-barium-copper-oxide tape. This novel magnet design produces an enormous magnetic field of more than 28 T—almost 600,000 times stronger than the Earth's magnetic field. The installation in Utrecht follows those at the University of Florence, ETH Zürich, and in Göttingen at the Max Planck Institute for Biophysical Chemistry. These powerful multi-million-dollar instruments will help to answer some of the most important questions facing our future such as studying the structure and function of proteins linked to diseases and viruses, including SARS-CoV-2.

These two instruments—MRI and NMR—represent the two main commercial uses of superconductivity.

The high magnetic fields possible with superconducting magnets have found application in nuclear fusion reactors and particle accelerators. Because of the enormous lengths of wire needed, these applications were initially only open to the low-temperature metallic alloy superconductors.

In the northwest suburbs of Geneva the Large Hadron Collider (LHC) at CERN built between 2002 and 2008 and ITER being constructed in Saint Paul-lez-Durance in southern France are flagship projects for superconducting wires. The Large Hadron Collider contains a total of 1200 tons of niobium titanium, which equates to around 7600 km of superconducting cable. The cable itself is made up of strands 0.825 mm in diameter. Each strand houses 6,300 filaments that are 0.006 mm thick (10 times thinner than a human hair.) If placed end-to-end the total length of the filaments would reach to the Sun and back five times with enough left over for a few trips to the Moon![22]

ITER is an ambitious global energy project involving thirty-five nations— China, the European Union, India, Japan, Korea, Russia, and the United States—in what has already been a more than 35-year collaboration to build the world's largest tokamak, a powerful magnetic fusion device designed to prove the feasibility of nuclear fusion as a large-scale and carbon-free source of energy. ITER models, but on a much smaller scale, the processes that happen within our Sun and the stars. If successful, the ITER project will produce 500 megawatts of fusion power from an input heating power of 50 megawatts— generating a ten-fold return.[23] ITER will use 100,000 km of helium-cooled niobium tin (Nb_3Sn) wire to generate the immense magnetic field that is necessary to confine the 150,000,000°C plasma at the heart of the tokamak. Ground was broken on the project in 2007 and the first plasma is scheduled for December 2025 with operation beginning in 2035. A total project time of fifty years—assuming all goes well.

Fusion is often promoted as an energy solution to rid us of our insatiable reliance on finite fossil fuels, a way to create a sustainable energy future eliminating harmful greenhouse gas emissions that contribute to global climate change. President Joseph R. Biden has pledged that the United States will eliminate all greenhouse gas emissions from its energy sector by an ambitious goal of 2035. (About the same time ITER may be up and running.) While reaching this goal will require a sharp increase in the use of wind and solar generation, nuclear fusion may eventually play a role if reactors such as ITER and others are successful.

Fusion reactors might also create a long-term market for high-temperature superconductors, which can produce higher magnetic fields leading to more compact reactors than those possible using metallic superconductors. This is the idea behind the SPARC fusion device, which is led by a team from the Massachusetts Institute of Technology's Plasma Science Fusion Center and MIT spin-out company, Commonwealth Fusion Systems. SPARC's magnets are wound using a complex multilayered cable that contains stacks of the yttrium-barium-copper-oxide superconductor.[24]

The large magnetic fields attainable with superconducting magnets raised exciting possibilities for a new form of high-speed travel using magnetic levitation. Levitating trains able to reach speeds well in excess of 370 mph would allow long distance train travel to compete with airplanes. For instance, a train journey from New York to Los Angeles could be completed in under seven hours. The idea of magnetically levitated or "maglev" trains was that of James Powell and Gordon Danby both researchers at Brookhaven National Laboratory in Upton, New York. They published the first paper on the concept of superconducting maglev trains in 1966 and were awarded a United States patent three years later, well before the discovery of high-temperature superconductors.[25]

Powell and Danby proposed using superconducting magnets carried on the train that would interact with electromagnets buried in the tracks. The repulsive force generated between the electromagnets and the superconductor would be sufficient to cause the train to levitate as much as 10 cm above the guideway. Magnetic fields would also be used to propel the train forward. Powell and Danby made a calculation that showed that a 100-foot-long passenger train weighing 60,000 pounds could be suspended magnetically using superconducting magnets. Their invention is the basis for the proposed Tokyo to Osaka maglev route designed to complete the 300 mile journey in about one hour. The Chuo Shinkansen line is planned to link Tokyo and Nagoya by 2027 in only forty minutes—faster than flying or using the existing train line. Extension of the maglev line from Nagoya to Osaka is expected by 2045. The superconducting coils will use a niobium titanium alloy cooled with liquid helium.

So, while low-temperature metallic superconductors based on niobium alloys have found some important, but limited, applications high-temperature superconductors are still looking for where their unique properties can be fully utilized. But they are not the only wonder material that has failed to live up to the early hype and even some of the later, more modest, expectations. The promise of the nanostructured forms of carbon for instance buckyballs and most notably carbon nanotubes remains at present largely unmet as commercial applications in cosmetics and as minor components in plastic composites are far from the outlandish application of using carbon nanotubes to form a sci-fi space elevator extending from the Earth thousands of miles into space.[26]

The major factor limiting the widespread use of high-temperature superconductors is the difficulty of forming them into the most useful shape—long thin wires. Complex designs combining superconductor-filled metal tubes—often made of silver—carefully wound around and inside special cooling channels for liquid nitrogen have added to the cost of using high-temperature superconductors. Limited cable lengths have made it very difficult to seriously consider spanning the enormous distances necessary for instance in power transmission.

Long thin wires are also the most useful form for winding superconductors into magnets. The large magnetic fields available with high-temperature superconductors is making them increasingly interesting for high field applications such as next generation NMR spectrometers. In 2021 a Japanese group led by the RIKEN Center for Biosystems Dynamics Research in Yokohama demonstrated an NMR magnet using high-temperature superconductors that could produce a magnetic field of more than 9 T.[27] One of the team's goals is to make a high-resolution 30.5 T magnet, but this might require combining both high-temperature and low-temperature superconductors.

Commercialization of carbon nanotubes is facing a similar challenge to that experienced by high-temperature superconductors: how to fabricate them into the most usable form. This is often as long thin wires or yarns assembled from enormous numbers of individual carbon nanotubes. The challenge is quite daunting: a space elevator of around 150,000 km in length might be composed of individual carbon nanotubes that are only 100 nm long. To date the longest single carbon nanotube was grown at Tsinghua University in Beijing, China and clocked in at half a meter long.[28] While impressive, this is tiny compared to the length of wires that can be drawn from metals.

Recently it was reported that making high-strength carbon nanotubes at all in a usable form is going to be difficult. Research led by the Institute of Textiles and Clothing at the Hong Kong Polytechnic University has shown that just a single out-of-place atom is enough the cut the strength of a single-walled carbon nanotube by more than half.[29] Theoretical studies have shown that a perfect single carbon nanotube can have a tensile strength of 100 gigapascals, making it one of the strongest materials around. To meet the requirements of a space elevator it has been calculated that cables with a tensile strength of 50 gigapascals would be required.[30] But efforts to spin multiple nanotubes into a practical large-scale fiber have only produced ropes with strengths of 1 gigapascal or less. The highest recorded strength in a very short carbon nanotube fiber is 9 gigapascals—a far cry from the theoretical value.[31]

It is not unusual for the actual strength of a material to be considerably less than that predicted by theory. In fact, that is always the case. English engineer Alan Arnold Griffith explained this discrepancy in 1920 while working at the Royal Aircraft Establishment in Farnborough. Griffith's theory was that preexisting flaws in materials act to concentrate stress. To prove his point, Griffith measured the tensile strength of a large number of glass tubes where a sharp crack had been deliberately introduced. What he found was that glass tubes with the larger cracks were weaker than those with smaller cracks. As the crack size increased the strength of the glass tube decreased.[32] Griffith also carried out an additional set of experiments using glass of the same composition—an English Hard Glass—where he measured the strength of glass fibers as a function of their diameter. The narrower fibers were found to be much stronger than the wider fibers—a 5 μm diameter glass fiber was four times stronger than one with a diameter of 20 μm. Griffith's explanation was that as the fibers got smaller, the probability of them containing a crack—a fatal flaw—decreases and for narrower fibers the size of the largest possible crack must also decrease. Consequently, they get stronger. The pioneering work of A.A. Griffith is the basis for the "weak link" statistical approach that was adopted by Swedish engineer Waloddi Weibull. In 1951 Weibull presented his paper to the American Society of Mechanical Engineers on a statistical model that could be used to assign a probability that a component would fail based on the theory that it will fail at its weakest point.[33] Hence there is a direct relationship between flaws in a material and the probability of failure.

For carbon nanotubes the large gap between their theoretical strength and those measured in practice is also due to the presence of defects often the result of alignment problems when a yarn is pulled from a carbon nanotube mat (sometimes referred to as a "forest".) Figure 9.2 shows two scanning electron microscope images of the same carbon nanotube yarn, at different magnifications, being drawn and twisted from a forest of multi-walled carbon nanotubes. Yarn lengths up to 1 m have been drawn using this method.

Even just a single atom out of place can weaken an entire carbon nanotube fiber—turning two of the hexagons into a pentagon and a heptagon, creating a kink in the tube as illustrated in Figure 9.3. One bad tube in the fiber can significantly reduce the strength of the final cable.

In what turned out to be an overly optimistic 2003 study funded by the NASA Institute for Advanced Concepts, Bradley Edwards suggested that an operational space elevator using carbon nanotube cables was possible within fifteen years at a cost of $10 billion (close to $15 billion in today's money.)[34]

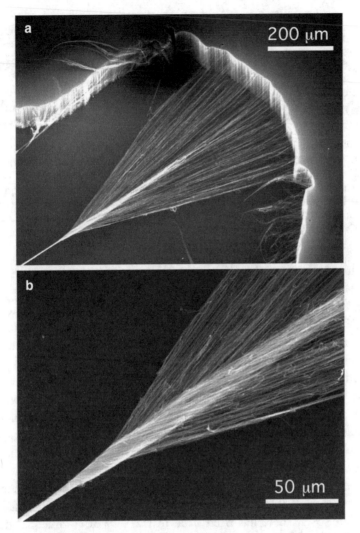

Fig. 9.2 Scanning electron microscope images of carbon nanotube yarns drawn from a carbon nanotube forest. b is a magnified image of a (Reprinted with permission from Zhang R, Zhaing Y, Zhang Q, Xie H, Qian W, Wei F (2013) Growth of half-meter long carbon nanotubes based on Schulz-Flory distribution. ACS Nano 7:6156–6161. Copyright 2013 American Chemical Society)

More conservative estimates at the time placed construction of what would likely be considered as a Wonder of the Modern World at between 50 and 300 years. We are thirty years since Sumio Iijima first observed carbon nanotubes using a powerful electron microscope. There is still a very long way to go before carbon nanotubes meet their promise of game-changing

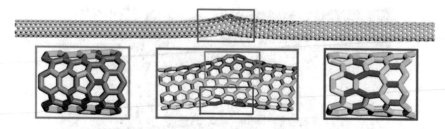

Fig. 9.3 Illustration of a single-walled carbon nanotube under tensile loading. One misplaced atom has created, in the left side image, a five-sided ring and on the right side image a seven membered ring. The result is a kink that creates a region of stress concentration (red color) that reduces the overall strength of the nanotube (Reprinted from Zhu L, Wang J, Ding F (2016) The great reduction of a carbon nanotube's mechanical performance by a few topological defects. ACS Nano 10:6410–6415. Copyright 2016 American Chemical Society)

real-world applications. Significant breakthroughs in synthesis are what is desperately needed.

Even thirty years after the discovery of carbon nanotubes it is still not possible to produce large quantities of high-quality material. Most synthesis runs produce a mixture consisting of single-walled and multi-walled nanotubes, buckyballs, soot, and other forms of carbon. These mixed powders can be added to polymers to form composite materials. Unfortunately, the results so far do not show that these additions provide any competitive advantage over more conventional carbon fibers. In addition to the challenges of producing a uniform product, carbon nanotubes when added to a polymer to form a composite tend to clump together creating a nonuniform dispersion that actually decreases strength and reduces performance.[35]

Graphene is another carbon nanomaterial that has generated an enormous amount of research interest and suggestions for an array of potential applications, but so far there has been little in the way of commercial use. But graphene may achieve what its derivative material, the carbon nanotube, has been unable to. As Edward Randviir and colleagues at Manchester Metropolitan University note there has been far more attention paid to graphene than carbon nanotubes, because of the unrivalled range of superior properties demonstrated by this one-atom layer thick material.[36] In particular, researchers look at the potential of graphene-based electronic devices one day replacing silicon as the material at the heart of all our consumer electronics and our interconnected world. But as Randviir and colleagues conclude: "The obstacles which need to be overcome are things such as mass production and graphene quality. For example, for many of the [proposed] applications to thrive they require large area, defect-free, grain boundary-free, monocrystalline graphene to be readily available and unfortunately, to date that has not been achieved."

Despite the lack of progress in commercialization of carbon nanotubes, graphene, and other nanostructured forms of carbon there are historical examples of where a "wonder" material discovered in a laboratory was able to make a successful transition from lab curiosity to multi-billion dollar product in less than a decade. Maybe there are useful lessons to be learnt. The example I will use is that of the polymer polytetrafluoroethylene (PTFE) that became the ubiquitous Teflon, which enjoys a current market size of around $3 billion. PTFE was the result of a fair dose of serendipity not unlike the discovery of buckyballs and carbon nanotubes.

It begins at the DuPont research laboratories in Deepwater, New Jersey. Researcher Roy J. Plunkett and his colleagues were attempting to replace the toxic chemicals, including sulfur dioxide and ammonia, that were used as coolants in refrigerators and air conditioners with what at the time were considered the more benign chlorofluorocarbons (CFCs). On the morning of April 6, 1938 Plunkett opened a cylinder that was supposed to be full of tetrafluoroethylene, a colorless, odorless gas. Surprisingly the cylinder released 990 g of gas rather than the full amount of 1000 g. Plunkett asked the question: what had happened to the missing 10 g. On cutting open the cylinder Plunkett found his answer. Inside was a powdery white residue, which when scraped out using a metal wire accounted for the missing material.[37] The tetrafluoroethylene had polymerized forming *poly*tetrafluoroethylene, which was disclosed in a United States patent in 1941 and later registered under the trade name Teflon®.[38]

Teflon's many superlative properties soon became evident: its chemical inertness, superior electrical insulation, heat and weather resistance, together with its amazingly low coefficient of friction. In a paper published by two DuPont chemists Malcolm Renfrew and E.E. Lewis in the journal *Industrial and Engineering Chemistry* in 1941 and presented at the American Chemical Society meeting in Atlantic City, New Jersey they noted that: "No substance has been found which will dissolve or even swell the polymer".[39] Renfrew and Lewis tested several hundred different reagents and found that "such potent reagents as aqua regia,… hot nitric acid, and boiling solutions of sodium hydroxide do not affect the polymer." Aqua regia is a powerful mixture of nitric acid and hydrochloric acid and notorious because of its ability to dissolve the precious metals gold and platinum.

Renfrew and Lewis were quite modest in their suggestion of applications for Teflon ranging from gaskets in pumps to pipes and tubes in chemical equipment. They further noted that wartime restrictions and scarcities of the material had led to limited exploration of possible industrial applications. Once the war was over (their paper was published in September 1946) they "anticipated that developments in this field now will proceed more actively." Little did they know, or speculate upon, the enormous range of applications that would be found for Teflon not only in chemical processing, but also in the automotive and transportation industries and in medicine where a porous form of Teflon called expanded PTFE (or ePTFE) is used by surgeons for artificial veins and arteries to treat cardiovascular disease. Once woven into a fabric ePTFE becomes Gore-Tex a breathable, waterproof fabric used in rainproof clothing products such as jackets manufactured by familiar brands including L.L. Bean, Patagonia, and The North Face.[40] The Gore company, the manufacturer of Gore-Tex, has annual sales of $3.8 billion and holds more than 3,400 patents for applications of ePTFE in areas from medical implants to electronics.

A process patent disclosing a method invented by Malcolm Renfrew to synthesize Teflon was filed by DuPont in November 1946 and issued in December 1950. In addition to describing its synthesis, the patent states that an important object is to provide such a process which is economical and relatively simple to carry out.[41] Production of Teflon as a commercial product, enabling applications in almost every industry sector, lagged only eight years behind its discovery by Roy Plunkett. Included among the processes that are used nowadays to commercially manufacture Teflon one is a version of the method originally patented by Renfrew.

Unlike the modest expectations set for Teflon or the caution shown by Bednorz and Müller in the tentative title of their paper, which won them a Nobel Prize, the tendency nowadays is to indulge in a little (or sometimes more) hyperbole concerning the impact of a particular study.[42] As materials scientists Andrew Bell at University of Leeds and Dragan Damjanovic at EPFL note there is an increasing tendency to describe the results of a study as "excellent" when that plaudit is not justified or to claim impressive commercial potential when factors such as a detailed cost–benefit analysis has not been completed.

Bell and Damjanovic suggest that the underlying reasons for this level of overoptimism are related to the importance given to the ranking of a journal where the studies are published, which can affect success in obtaining a job, the outcome of promotion and tenure cases, or in deciding promotions and pay rises. Overstating the potential socioeconomic impact of the research helps justify publication in leading journals and thereby increasing the academic impact of the paper. Universities and publishers benefit because it increases rankings and reputation and for publishers leads to more downloads and citations. Bell and Damjanovic pose the question: "Can the exaggerated claims for the application potential of a material be detrimental?" The answer they give is, "Where hyperbole becomes an issue is when non-experts are making decisions or recommendations based on information they read in the scientific literature." With increasing levels of public opinion differing from that of scientific experts and with increasing levels of popular distrust of expert opinion making even mildly exaggerated claims in scientific publications can have "wider consequences and may contribute to the continued erosion of public trust in publicly funded science."

Notes

1. The Nobel Prize in Physics 1987 NobelPrize.org. Nobel Media AB 2021. Mon. 26 Apr 2021. https://www.nobelprize.org/prizes/physics/1987/summary/. The award was made just one year after publication of the discovery of high-temperature superconductivity.
2. Bednorz JG, Müller KA (1986) Possible high Tc superconductivity in the Ba-La-Cu-O system. Zeitschrift für Physik B-Condensed Matter 64:189-193. This is a classic paper with a cautionary title. It has now been cited almost 20,000 times since its publication.
3. Michel C, Er-Rakho L, Raveau B (1985) The oxygen defect perovskite $BaLa_4Cu_5O_{13.4}$, a metallic conductor. Materials Research Bulletin 20:667–671.
4. Bardeen J, Cooper LN, Schrieffer JR (1957) Theory of superconductivity. Physical Review 108:1175–1204. The 1972 Nobel Prize in Physics was awarded jointly to John Bardeen, Leon Neil Cooper, and John Robert Schrieffer "for their jointly developed theory of superconductivity, usually called the BCS theory. Bardeen won the Prize twice—1956 and 1972.
5. Most studies of high-temperature superconductivity indicate the presence of Cooper pairs below T_c. The discussion in the physics community seems more about how the pairing happens. A recent study suggests that paramagnons might be prominent candidates for mediating Cooper pairing in high-temperature superconductors: Chu H, Kim MJ, Katsumi K, Kovalev S et al.(2020) Phase-resolved Higgs response in superconducting cuprates. Nature Communications 11:1793.
6. If a material glows dark red, it is at a temperature between 700–800°C. When bright white it is at 1400°C.
7. The Meissner effect is named after German physicist Walther Meissner who together with Robert Ochsenfeld discovered the phenomenon in 1933.
8. U.S. Congress, Office of Technology Assessment, High-Temperature Superconductivity in Perspective, OTA-E-440 (Washington, DC: U.S. Government Printing Office, April 1990) p 126.

9. Willls I (2019) The Edisonian method: trial and error. In: Thomas Edison: Success and Innovation through Failure. Studies in History and Philosophy of Science, vol 52. Springer, Cham. Wills quotes Nicola Tesla, who worked for Edison from 1882 to 1883, as saying: "His method was inefficient in the extreme, for an immense ground had to be covered to get anything at all unless blind chance intervened and, at first, I was almost a sorry witness of his doings, knowing that just a little theory and calculation would have saved him 90% of the labour (sic)." The trial-and-error method that Tesla derided became so closely associated with Thomas Edison that it is often referred to as the Edisonian method.

10. Tinkham M, Beasley MR, Larbalestier DC, Clark AF, Finnemore DK (1988) Workshop on Problems in Superconductivity, Copper Mountain, Colorado, p 12.

11. Wu MK, Ashburn JR, Torng, CJ, Hor PH, Meng RL, Gao L, Huang ZJ, Wang YQ, Chu CW (1987) Superconductivity at 93-K in a new mixed-phase Y-Ba-Cu–O compound system at ambient pressure. Physical Review Letters 58:908–910.

12. Chu CW, Gao L, Chen F, Huang ZJ, Meng RL, Xue YY (1993) Superconductivity above 150 K in $HgBa_2Ca_2Cu_3O_{8+d}$ at high pressures Nature 365:323–325.

13. Drozdov AP, Kong PP, Minkov VS et al. (2019) Superconductivity at 250 K in lanthanum hydride under high pressures Nature 569:528–531.

14. Jones EG, Lineweaver CH (2010) Pressure–temperature phase diagram of the Earth. Pathways Towards Habitable Planets, ASP Conference Series 430:145–151.

15. Snider E, Dasenbrock-Gammon N, McBride R et al.(2020) Room-temperature superconductivity in a carbonaceous sulfur hydride. Nature 586:373–377.

16. In addition to superconductors having a critical temperature, T_c, there is also a critical current density, J_c. A current density greater than J_c will cause a loss in superconducting properties.

17. Oestergaard J (1997) Superconducting power cables in Denmark—A case study. IEEE Transactions on Applied Superconductivity 7:719–722.

18. Uglietti, D (2019) A review of commercial high temperature superconducting materials for large magnets: from wires and tapes to cables and conductors Superconductor Science and Technology 2: 053,001.

19. The Nobel Prize in Physics (1913) NobelPrize.org. Nobel Media AB 2021. Tue. 27 Apr 2021. https://www.nobelprize.org/prizes/physics/1913/summary/ >.

20. Parizh M, Lvovsky Y, Sumption M (2017) Conductors for commercial MRI magnets beyond NbTi: requirements and challenges. Superconductor Science and Technology 30:014,007.

21. The Nobel Prize in Physics (1952) NobelPrize.org. Nobel Media AB 2021. Wed. 28 Apr 2021. <https://www.nobelprize.org/prizes/physics/1952/sum mary/ .

22. http://lhc-machine-outreach.web.cern.ch/components/cable.htm.

23. This value is similar to the generating capacity of a single fission reactor at the Prairie Island nuclear plant in Minnesota, which has two 520 MW reactors. The data comes from the U.S. Energy Information Administration, https:// www.eia.gov/tools/faqs/faq.php?id=104&t=3.

24. Hartwig ZS, Viera RF, Sorbom BN et al. (2020) VIPER: an industrially scalable high-current high-temperature superconductor cable. Superconductor Science and Technology 33:11LT01.

25. Powell JR, Danby GR (1966) ASME Railroad Division Annual Meeting, New York, Report No. 66-WA/RR-5. A United States Patent 3,470,828 for Electromagnetic inductive suspension and stablilization system for a ground vehicle was filed November 21, 1967 and issued October 7, 1969.

26. Pugno NM (2006) On the strength of the carbon nanotube-based space elevator cable: from nanomechanics to megamechanics. Journal of Physics: Condensed Matter 18: S1971-S1990.

27. Yanagisawa Y, Piao R, Suetomi Y, Yamazaki T, Yamagishi K et al. (2021) Development of a persistent-mode NMRE magnet with superconducting joints between high-temperature superconductors. Superconductor Science and Technology 34:115,006.

28. Zhang R, Zhang Y, Zhang Q, Xie H et al. (2013) Growth of half-meter long carbon nanotubes based on Schulz-Flory distribution. ACS Nano 7:6156–6161.

29. Zhu L, Wang J, Ding F (2016) The great reduction of a carbon nanotube's mechanical performance by a few topological defects. ACS Nano 10:6410–6415.

30. Yakobson BI, Smalley RE (1997) Fullerene nanotubes: $C_{1,000,000}$ and beyond. American Scientist 85:324–337. Richard Smalley shared the 1996 Nobel Prize in Chemistry for the discovery of fullerenes.

31. Koziol K Vilatela J, Moisala A et al. (2007) A high-performance carbon nanotube fiber. Science 318:1892–1895.

32. Griffith AA (1921) The phenomena of rupture and flow in solids. Philosophical Transactions of the Royal Society A 221:582–593. The paper was read February 29, 1920 and published January 1, 1921. Not only was Griffith involved in fundamental studies of the properties of materials he also made important contributions in many areas of aircraft design including developing design rules for air-cooled engines.

33. Weibull W (1951) A statistical distribution function of wide applicability. ASME Journal of Applied Mechanics September: 293–297. Although Weibull applied his model to the failure probability of materials it actually has a broad applicability. One additional example in Weibull's original paper

was applying the model to the statures for adult males born in the British Isles.

34. Edwards BC (2003) The space elevator. NIAC Phase II Final Report, March 1. Available at http://www.niac.usra.edu/files/studies/final_report/521 Edwards.pdf.

35. Kinloch IA, Suhr J, Lou J, Young RJ, Ajayan PM (2018) Composites with carbon nanotubes and graphene: An outlook. Science 362:547–553.

36. Randviir, Edward P, Brownson, Dale AC, and Banks, Craig E (2014) A decade of graphene research: production, applications and outlook. Materials Today 17:426–432.

37. This description of the discovery of Teflon is as described by Emsley J (1998) Molecules at an Exhibition. Oxford University Press, Oxford, p 132 and is supported by Plunkett's personal communication to Garrett AB (1962) The Flash of Genius, 2 Teflon: Roy J. Plunkett. Journal of Chemical Education 39:288.

38. Plunkett RJ (1941) U.S. Patent 2,230,654.

39. Renfrew MM, Lewis EE (1946) Polytetrafluoroethylene: heat-resistant, chemically inert plastic. Industrial and Engineering Chemistry 38:870–877. Renfrew and Lewis worked at DuPont's Arlington New Jersey facility, which closed in 1959.

40. Gore RW, Allen Jr, SB (1978) Waterproof laminate. U.S. Patent 4,194,041.

41. Renfrew MM (1950) Polymerization of tetrafluoroethylene with dibasic acid peroxide catalysts. US Patent 2,534,058. The patent application was filed in November 1946 and issued in December 1950. Renfrew Hall at the University of Idaho is named in honor of Malcolm Renfrew where he served as head of chemistry. Malcom Renfrew passed away at the age of 103.

42. Bell AJ, Damjanovic D (2020) Balancing hyperbole and impact in research communications related to lead-free piezoelectric materials. Journal of Materials Science 55:10,971–10,974.

10

A Green New Deal

A remarkable study led by the University of California Santa Barbara looked at the production, use, and fate of all plastics ever made.[1] The starting point for the study was taken as 1950 when plastic production totaled a mere 1.5 million tons.[2] Two decades later, annual world production of plastics had reached 27 million tons exceeding that of both aluminum and copper combined.[3] In the sixty-five years between 1950 and 2015—the period of the study—global production of plastics amounted to a staggering 8,300 million tons or 8.3 billion tons. That is equivalent to the weight of over 22,000 Empire State Buildings. From this amount of primary plastic production, 6,300 million tons of plastic waste was generated. The researchers found that only 9% of the waste had been recycled, 12% was incinerated, and the vast majority, almost 80%, either accumulated in landfills or found its way into the natural environment. Plastic waste is now present in rivers and streams, in oceans and seas, and in almost every landmass that has been studied.

Plastics make up about 90% of marine litter. The Great Pacific Garbage Patch is a collection of marine debris in the North Pacific Ocean located between Hawaii and California formed by rotating ocean currents called "gyres." As they swirl, like giant whirlpools, gyres draw in litter, discarded fishing gear, and other ocean debris to form large amorphous floating masses of trash that cover areas of thousands of square miles and reach from the surface of the water all the way to the ocean floor. There are five ocean gyres. In addition to two in the Pacific Ocean—one in the western part of the ocean,

© The Author(s), under exclusive license to Springer Nature Switzerland AG 2023
M. G. Norton, *A Modern History of Materials*,
https://doi.org/10.1007/978-3-031-23990-8_10

the other in the east—the Atlantic Ocean also has two and there is one gyre in the Indian Ocean.

Perhaps the most well known of the five, the Great Pacific Garbage Patch consists of every imaginable form of plastic from discarded fishing nets to medical waste to children's toys to plastic containers, to tiny microplastics, pieces of plastic smaller than 5 mm in diameter. While the larger pieces of debris, for instance the plastic bottles and plastic bags are easily spotted, the microplastics are much less obvious and can be difficult to see, particularly if they are below the surface of the water. Despite various efforts to round up and dispose of the waste, the Great Pacific Garbage Patch grows at a rate of 1 to 2 million tons of plastic waste per year. That is equivalent to over 100 billion ½-liter soda bottles.

The environmental impact of marine debris is easy to see and has been documented by the United States National Oceanic and Atmospheric Administration (NOAA).[4] NOAA categorized environmental impact into three categories—entanglement and ghost fishing, ingestion, and non-native species. Ghost fishing refers to discarded nets continuing to catch fish even though they are no longer under the control of the fisher. Ghost nets not only trap fish they frequently entangle other species for instance sea turtles, dolphins, and seals. Once trapped it is almost impossible for the animal to break free even with the aid of divers. Animals may mistakenly eat plastic and other debris. We know that this can be harmful to the health of fish, seabirds, and other marine animals. Pictures of the stomachs of dead seabirds show among the many unidentifiable pieces of plastic, larger items such as small containers, bottle tops, and even disposable cigarette lighters. These items take up room in their stomachs, making the animals feel full and stopping them from eating real food. And finally, as reported by NOAA marine debris can transport species from one place to another. Algae, barnacles, crabs, or other species can attach themselves to debris and be transported across the ocean. If these species are invasive, and can settle and establish in a new environment, they can compete with or overcrowd native species, disrupting an entire ecosystem.

How did all this marine debris get there?

It is estimated that as much as 2.41 million tons of plastic waste currently enters the world's oceans every year from rivers.[5] China's Chang Jiang (Yangtze River) which flows 3,900 miles from Jari Hill in the Tanggula Mountains to the East China Sea was found to be the worst culprit, responsible for 1.47 million tons of plastic pollution reaching the oceans in 2017 alone.[6] This waste is just part of the between 15 and 51 trillion plastic particles floating in and on the world's oceans.

A study by a team of scientists from Bangor University in Wales found microplastics in every sample they collected from bodies of water across the length and breadth of Great Britain. The scenic waters of Loch Lomond in Scotland had 2.4 pieces of microplastic per liter, making it the least contaminated of the waterways studied. The River Thames, which rises in Gloucestershire and flows through the heart of London before flowing to the North Sea contained 84.1 pieces of microplastic per liter. But by far the most contaminated waterway was the River Tame in Greater Manchester where over 1,000 pieces of microplastic were recorded per liter of water. The researchers stopped counting when they reached 1,000![7] What was the source of this large quantity of microplastics? Untreated wastewater.

Careless waste disposal and inefficient collection cause plastic waste to build up in cities across the world from New York City to Rio de Janeiro and from London to Sydney. Even the once pristine environment of Mount Everest, the world's highest peak, is not immune from contamination by plastic waste accumulating rapidly from the increased tourism on the mountain. Almost 900 people climbed Everest in 2019. That same year a National Geographic expedition assessed the extent of microplastic pollution on Everest. Led by Imogen Napper of Plymouth University in the United Kingdom the scientists found an average of 30 microplastic particles per liter of water in the snow samples they examined. In the most contaminated sample, collected at an elevation of 8,440 m above sea level, they counted 119 particles per liter.[8]

The plastic waste on Mount Everest includes the highly visible tents and climbing ropes and the much less obvious microfibers shed from the "Himalayan suits" made of waterproof polyester, acrylic and polypropylene fibers. Ultra-violet light combined with mechanical abrasion wear down the discarded plastic waste into smaller and smaller particles until they become the ubiquitous microplastics. These tiny plastic fragments can be blown long distances by the wind and carried by streams dispersing them far and wide from their starting point.

We are unfortunately familiar with the sight of an oil tanker spilling its load, which makes its way to the shore where it can kill many hundreds of seabirds and damage a fragile coastal ecosystem. But we are perhaps less familiar with the concept of a "plastic spill." In May 2021 that changed when the Singapore-flagged MV X-Press Pearl cargo ship caught fire and sank off Sri Lanka's west coast. Among the toxic mass of charred fuel oil and burnt debris that ended up on Sarakkuwa beach polluting the coastline for hundreds of miles in each direction were millions of tiny plastic pellets, nurdles. Nurdles are the small pea-sized plastic pellets, larger in size than microplastics, that are the raw material used in the manufacture of a whole host of plastic products including containers, bottles, and as mentioned in Chapter 4, polyethylene films. A report of the United Nations Environmental Advisory Mission not only labelled the incident as the worst maritime disaster to have struck Sri Lanka, but also the "single largest plastic spill" in history with over 1,600 tons of plastic released into the Laccadive Sea.[9]

In a recent study conducted jointly by the School of Atmospheric Sciences at Nanjing University and the Scripps Institution of Oceanography at the University of California San Diego it was found that, perhaps not unexpectedly, the recent COVID-19 pandemic has led to an increased demand for single-use plastic. According to work done by the team of researchers more than eight million tons of pandemic-associated plastic waste have been generated globally, with more than 25,000 tons entering the world's oceans. Most of the plastic was found to be from medical waste from hospitals.[10] Smaller amounts were from personal protection equipment including gloves and masks and packaging materials from increased volumes of online shopping.

If 8,300 million tons of plastic seems a very large number, and it is, then the projection for the cumulative amount of plastic waste generated by the middle of the century is truly staggering. Assuming production continues at the same rate Roland Geyer, Jenna Jambeck, and Kara Law, the three co-authors of the University of California Santa Barbara-led study, project that by 2050 the amount of plastic waste will exceed 25,000 million tons. Assuming consistent use patterns and projecting current global waste management trends to 2050, 9000 million tons of plastic waste will have been recycled, 12,000 million tons incinerated, and 12,000 million discarded.

Even though the projections indicate that over one-third of the plastics produced will be recycled, they were eventually destined to enter the waste stream to be either incinerated or find their way into landfills or contaminate the natural environment including our oceans and streams.

Our present relationship with plastics follows the linear materials approach of *take—make—use—dispose* described by Cambridge University's Michael Ashby that goes from natural resource, through product, to landfill.[11] We can illustrate each of these steps with examples of specific plastics.

take—This refers to the raw materials used in the manufacture of a product, including how and where we obtain them. Since the invention of Bakelite, the first fully synthetic plastic, by American chemist Leo Hendrick Baekeland in 1907 virtually all the plastics we use on a daily basis are derived from fossil fuel hydrocarbons. Bakelite is what is known as a phenol–formaldehyde resin. The two components, phenol and formaldehyde, are derived from coal tar and methanol, respectively. Coal tar is a liquid produced when coal is heated to temperatures between 900°C and 1200°C in the absence of air. Phenol is a liquid fraction within the tar that can be separated from the other fractions by collecting the component that boils in the temperature range 170 to 230°C. Although the current industrial production of phenol does not involve coal tar it is still synthesized using chemicals derived from fossil fuels. At the time of Baekeland's invention methanol, commonly known as wood alcohol, was obtained by the destructive distillation of wood. The process involves decomposing wood at a temperature between 450°C and 550°C then separating out the methanol by condensation. Although wood is no longer the raw material used in the synthesis of methanol it does rely on natural gas, a fossil fuel, as the carbon source. Our most widely used of all plastics is polyethylene from which we make everything from milk jugs to grocery bags. Almost all our liquids whether bleach, laundry detergent, or shampoo are contained in polyethylene. Industrial manufacture of polyethylene uses ethylene that is obtained from petroleum or natural gas, both fossil hydrocarbons.

In total fossil fuels represent 99% of the base raw materials for the manufacture of plastics. About 4% of the world's fossil fuel resources are used in plastics production.

make—This refers to the processes that turn the raw material into a usable product, which can often involve purification followed by shaping. There are two main processes used to produce plastics—polymerization and polycondensation—and they both require specific catalysts to facilitate the reaction. In a polymerization reactor, monomers for instance ethylene and propylene are linked together to form long molecular chains. These are the macromolecules mentioned in Chapter 4, which were described by German chemist Hermann Staudinger. Because the ethylene molecule is so stable the synthesis of polyethylene requires the use of a catalyst to help break the strong carbon to carbon bond. The most common catalysts are based on titanium, for instance titanium chloride, which are also used in combination with organoaluminum compounds including triethylaluminum ($Al(C_2H_5)_3$.) These are called Ziegler–Natta catalysts after chemists Karl Ziegler and Giulio Natta. Ziegler and Natta would share the 1963 Nobel Prize in Chemistry "for their discoveries in the field of the chemistry and technology of high polymers."[12] In 1953 while Director of the Max-Planck-Institute für Kohlenforschung Karl Ziegler developed a method for creating long molecular chains from small molecules using aluminum compounds as catalysts. The role of the catalyst was to speed up the reaction and allow it to happen at more manageable—and commercially viable—conditions of temperature and pressure. With financial support from Montecatini, a large Italian chemical company

Giulio Natta built upon Ziegler's research and in 1955 discovered a catalyst that formed molecular chains that had very specific orientations and shapes thereby creating a new class of complex polymers with unique properties.

For high-density polyethylene (HDPE, resin code 2), the type used in containers such as milk jugs and detergent bottles, the polymerization process is carried out at pressures between 10 and 80 atmospheres and temperatures around 100°C. For low-density polyethylene (LDPE, resin code 4), the form used for plastic food wrap and electrical insulation much higher pressures are required in the range 1000 to 3000 atmospheres and temperatures as high as 300°C. The pressures for low-density polyethylene synthesis are very high and would be in the range of pressures encountered 2 to 6 miles below the Earth's surface.

Because the processes used to produce most plastics require both elevated temperatures and high pressures, there is an embodied energy associated with their manufacture. In other words, a certain amount of energy must be used to convert the raw material into the final product. A goal for a green, sustainable, economy would be to use materials that have low embodied energies. For most plastics embodied energies are in the range 60 to 120 megajoules per kilogram.[13] As a specific illustration, Bakelite has an embodied energy of 79 megajoules per kilogram. This is similar to that for polyethylene, which has an embodied energy of 81 megajoules per kilogram. We can compare these values with those for other widely used container materials. For example, aluminum has a very high embodied energy of over 200 megajoules per kilogram, which reflects the large amount of energy required to extract and purify bauxite, the main ore of aluminum, and then to use high temperature electrolysis to separate out the metal. On the other hand, a typical container glass has an embodied energy of just 11 megajoules per kilogram. This value can be further reduced if recycled glass is added to the mix in addition to new virgin raw materials.

use—Here we can give a single example of the many uses for plastics, the plastic grocery bag. In recent years the ubiquitous single-use plastic bag has been vilified as an example of our disposable, and unsustainable, economy. Most grocery bags are made from high-density polyethylene. American company Waste Management, Inc. estimates that four trillion plastic bags are used each year worldwide. Fourteen billion of those are used in the United States requiring the consumption of 12 million barrels of oil.

In 2002, Ireland became the first country to enforce a plastic bag tax, the "Plastax." This 15-cent fee (raised to 22 cents in 2007) was initiated in response to the country's annual consumption of 1.2 billion shopping bags and its growing plastic pollution problem. Plastax was also designed to eliminate one of the country's biggest imports since less than a quarter of the plastic bags used were manufactured domestically. There was an immediate effect of the tax on consumer behavior. Within the first few weeks, plastic carrier bag consumption fell by 94%, decreasing Ireland's annual consumption by about 1 billion bags. During its first year in place $9.6 million was raised from the tax, which was managed by the Environmental Fund and used to help finance various environmental initiatives including education and training programs.

Following Ireland's example 170 nations pledged to significantly reduce the use of plastic by 2030. Kenya and the United Kingdom are two examples. Kenya banned single-use plastic bags in 2017 and in June 2020 prohibited visitors from taking single-use plastics including water bottles into national parks, forests, beaches, and conservation areas. The United Kingdom introduced a tax on plastic bags in 2015 and banned the sale of products containing microbeads, including certain shower gels and face scrubs, in 2018. In the United States there is no federal ban, although several states including New York, California, and Hawaii have banned single-use plastics.

Despite the success of many of these programs in reducing the use of *plastic* bags, the number of *paper* bags has skyrocketed during the same period. This has resulted in some unintended trade-offs as found in a 2018 study by the Danish Environmental Protection Agency, which completed a very detailed cradle-to-grave examination of the various types of grocery bag available in Danish supermarkets.[14] The goal of the study was to determine which type of carrier bag had the lowest environmental impact considering a wide range of indictors from climate change to ecosystem toxicity to water resource depletion. Perhaps rather surprisingly low-density polyethylene (the type of polyethylene examined in the Danish study) was determined to have the lowest impact in nine of the fifteen categories. Unbleached paper bags were found to have the lowest impact in the three categories of climate change, cancer causing, and fossil fuel depletion. A further calculation was made to determine the number of primary reuse times for each bag type to provide the same environmental performance as the average low-density polyethylene carrier bag. Comparing all the environmental indicators an unbleached paper bag would have to be reused 43 times. A striking outcome from the study was that an organic cotton reusable bag would have to be reused 20,000 times when considering all indictors to provide the same environmental performance as the average low-density polyethylene carrier bag!

dispose—This is the final step in the life of a material following our linear model. As a specific example, let's consider the short life of a plastic water bottle. In 2020 the sales of bottled water in the United States reached 15 billion gallons. Worldwide bottled water is a $240 billion market growing at a projected compound annual growth rate of 11% to reach over $500 billion by 2028. The type of plastic most commonly used to make water bottles is polyethylene terephthalate (PETE or PET, resin code 1). It is clear, strong, and lightweight and can be easily formed into almost any shape using low-cost extrusion or molding processes. Although polyethylene terephthalate is completely recyclable and can be turned into new bottles the current recycling rate for polyethylene terephthalate in the United States is around 30%, with the remainder going to landfill, entering the natural environment, or being incinerated.[15] In Europe the recycling numbers for polyethylene terephthalate are better at over 50%, but that still leaves a considerable amount of material that is wasted.

Like most of our plastics, polyethylene terephthalate is not biodegradable and takes an estimated 1,000 years to decompose in nature. So simply discarding polyethylene terephthalate and other plastics does not represent a viable long-term solution to dealing with our increasing consumption of plastic products. Incineration in power plants to generate electricity uses the energy value in plastics, after all they are mainly carbon and hydrogen. But incineration comes with its own set of problems. These are mainly due to the generation of toxic fumes that contain carcinogenic substances including dioxins and polychlorinated biphenyls (PCBs), the generation of greenhouse gases (for example carbon dioxide) and left over resides similar to fly ash from coal-fired power plants that also contain toxins that must be disposed of extremely carefully so that they do not enter the natural environment. Because of the limitations of other methods of plastic disposal recycling is often considered the best option to manage polyethylene terephthalate waste.

So, moving away from the linear model of *take-make-use-dispose*, which is clearly not sustainable, what are some of the challenges of recycling as the last step in the life of a product and why are many recycling rates so low?

To understand why the recycling rates of polyethylene terephthalate are so low it is instructive to look at the way the material is recycled. After the polyethylene terephthalate products are separated from other components in the waste stream they are washed, ground, and crushed into nurdles, which can be the raw material for reprocessing into new products. Unfortunately, contamination of the waste stream by different types of plastic can result in poor quality products. For instance, a water bottle may be made of polyethylene terephthalate, but the cap is frequently made of polypropylene—a very different plastic to polyethylene terephthalate having a much lower melting temperature. This method of mechanical recycling—grinding and crushing—is mainly used for producing low-grade plastic products. So called "downcycling." A water bottle, for example, becomes the fiber insulation in a winter coat, but rarely is it formed into a new bottle.

Contamination is not the only factor that limits recycling of plastics. Recycling a polymer multiple times produces a change in the molecular structure of the material causing it to become weaker. A study investigating the closed-loop recycling of high-density polyethylene found that after 20 cycles through an extrusion process the length of the molecular chains comprising the plastic had decreased by 20%. Over that same number of recycles the strength of the polymer had decreased by almost 30%.[16] So even for controlling for all sources of contamination there is a physical limit on the number of times current polymers can be recycled into an identical product.

Downcycling while producing useful products has only delayed, but not prevented, the material going to a landfill or other type of disposal.

An alternative to mechanical recycling is to use chemicals that can break down the polyethylene terephthalate into smaller constituent molecules—the monomers—that can be reassembled into brand new polymer resin. This process is often referred to as depolymerization and is one half of a polymerize-depolymerize cycle that could keep discarded plastic out of the environment. At the present time chemical recycling processes are neither environmentally friendly nor are they cost effective due to the large quantities of chemicals required. Although chemical recycling of polyethylene terephtahlate is not widely practiced in 2019 oil and gas giant BP created BP Infinia with a technology to depolymerize polyethylene terephthalate waste into new feedstock that could be the raw material to make new plastic products.

None of the commonly used plastics are biodegradable. Why? The reason is quite simple. These materials, with their enormously long carbon chains that may comprise more than 100 carbon atoms, do not exist in nature, and therefore, there are no naturally occurring organisms that can break them down effectively or at all. Or are there?

In 2017 a team of scientists from Cambridge University in the United Kingdom and the Universidad de Cantabria in Spain reported the fast biodegradation of polyethylene by larvae of the wax moth *Galleria mellonella*, producing ethylene glycol.[17] Ethylene glycol is an extremely useful chemical. It is a valuable raw material for the manufacture of polyester fibers and polyethylene terephthalate as well as being an important ingredient in antifreeze solutions.

In their experiment the researchers placed a polyethylene grocery bag, apparently obtained from a Marks and Spencer store, in direct contact with wax worms. After about 40 minutes they noticed holes starting to appear in the polyethylene film. After leaving 100 wax worms, plus or minus, in contact with the polyethylene bag for half a day the bag had lost 92 mg. As a useful benchmark a typical polyethylene shopping bag weighs around 6 g. To determine whether the bag was actually being chewed by the wax worms or whether there was a chemical reaction that was responsible for weight loss the researchers smeared onto the plastic film a paste consisting of ground up wax worms. This experiment confirmed that breakdown of the plastic was due to a chemical reaction rather than the mechanical action of masticating wax worms. The average hourly degradation rate was calculated to be 0.23 mg per square centimeter.

In a separate study reported in the journal *Science*, a group of researchers from Japan demonstrated that the bacterium *Ideonella sakaiensis* 201-F6 could degrade polyethylene terephthalate film.[18] When grown on polyethylene terephthalate this bacterium produced enzymes, which converted the polymer into its two raw materials, terephthalic acid and ethylene glycol. Over a period of 80 days the film lost 60 mg.

Converting waste plastic back into raw materials, whether value-added chemicals like ethylene glycol or monomers that can be converted back into new plastics, is an example of "upcycling", which offers considerable benefits over downcycling. Upcycling enables the possibility of moving towards a more sustainable circular economy because the chemical value of the polymer—all those carbon and hydrogen atoms bonded together—is not lost.

Another recent approach to upcycling is to convert waste plastic into value-added high-quality liquids and waxes. As an illustration, a research team led by scientists at Argonne National Laboratory in Illinois demonstrated how to convert single-use polyethylene into motor oil.[19] The reaction was carried out at a temperature of 300°C under a hydrogen pressure of 170 pounds per square inch (more than ten times atmospheric pressure) in the presence of a catalyst. Figure 10.1 is a transmission electron microscope image of the catalyst, which consists of platinum nanoparticles dispersed on the surface of perovskite nanocuboids.

The perovskite nanocubes are up to 100 nm across and are decorated with platinum nanoparticles that are only about 12 atoms across. After a reaction time of 96 h 50 mg of polyethylene had been upcycled into motor oil and other hydrocarbon liquids. The conditions used in this particular reaction will likely make scale-up a significant challenge, not least because of the requirement for the use of a very expensive platinum catalyst. But this research provides a useful demonstration of how to utilize the chemical value of what is widely regarded as a waste material.

Fig. 10.1 Electron microscope images of platinum nanoparticles (light contrast) with an average size of 2 nm deposited on perovskite nanocuboids. The histogram in the left-hand size image shows the distribution of sizes of the platinum nanoparticles. Some are as small as 1 nm across (Reprinted from Celik G, Kennedy RM, Hackler RA, Ferrandon M, Tennakoon A et al (2019) Upcycling single-use polyethylene into high-quality liquid products. ACS Central Science 5:1795–1803 https://pubs.acs.org/doi/10.1021/acscentsci.9b00722. Further permissions related to the material excerpted should be directed to the ACS)

The approximately 300 million tons of single use plastic that are disposed of each year represent an enormous carbon-containing resource for the production of value-added chemicals and as the raw materials for the manufacture of new materials. According to the paper's authors if efficient technologies could be developed for extracting the value in waste polymers it would be equivalent to recovering about 3.5 billion barrels of oil each year.

As mentioned at the beginning of this chapter, by far the predominant resource model in today's globalized society is the linear *take—make—use—dispose*. This model not only applies to plastics. In fact, it reflects our relationship with many other types of material. According to Imperial College London's Transition to Zero Pollution Programme the world economy consumes approximately 100 billion tons of raw materials a year—mainly newly extracted minerals, metals, fossil fuels, and biomass. About half goes into making long-lasting products for instance infrastructure, cars, and heavy machinery. The other half is used in the production of shorter-lived products such as food, clothing, and plastics, which typically have a lifespan of a year or less before they are disposed of. Less than 10% of this waste is recycled. The rest goes to landfill, the incinerator, or is dumped into the environment.

A goal of sustainable development is that rather than "*take—make—use—dispose*", which is the path followed by around 90% of the materials we use, that there is instead a circular economy where once a product reaches the end of its life that the materials are reused in the production of new products. At the same time reducing our need to "take". A plastic water bottle is broken down into its original polymers and a new bottle, indistinguishable from the original is created. With no need to dip again into the finite supply of fossil fuels.

While achieving a circular economy is an essential goal that has been expressed in recent years by many countries, there is evidence that the concept of recycling and reuse of waste materials is not new. In 2020 an international team of archeologists working in the city of Italian city of Pompeii, which in AD 79 was buried under a thick carpet of volcanic ash when Mount Vesuvius erupted, have found huge mounds of refuse containing lumps of mortar and plaster, which could be repurposed as materials for the construction of new buildings. So, two thousand years ago in the ancient city of Pompeii the Romans had developed one of the first examples of a circular economy. Allison Emmerson an anthropologist at Tulane University in New Orleans describes what the researchers found:

> We found that part of the city was built out of trash. The piles outside the walls weren't material that had been dumped to get rid of it. They're outside the walls being collected and sorted to be resold inside the walls. The Pompeiians lived much closer to their garbage than most of us would find acceptable, not because the city lacked infrastructure and they didn't bother to manage trash, but because their systems of urban management were organized around different principles. This point has relevance for the modern garbage crisis. The countries that most effectively manage their waste have applied a version of the ancient model, prioritizing commodification rather than simple removal.[20]

An essential step as the Romans appreciated was recycling. Without efficient and extensive recycling programs a circular economy will not be achievable. It is not just plastics that have proven difficult to recycle efficiently and effectively. Many of our critical minerals end up in landfills or polluting the environment because economically viable recycling programs have not been developed. This is particularly true for the many elements that make up electric vehicles. As we mentioned in Chapter 6 a study by the International Monetary Fund projected that by 2042 over 90% of the vehicles on the road could be electric. And it is not just cars that are being electrified. There are now electric trucks, buses, motorbikes, bicycles, scooters, ships, and even more electric airplanes. To ensure a transition towards a circular economy, when all these machines come to the end of their useful life their constituent materials will need to be recycled. A challenge that recyclers are facing with electric vehicles compared with their gasoline or diesel counterparts is in the wide variety of materials that are used in electric vehicles. In particular, the numerous materials that are the active components inside car batteries and electric motors.

In a traditional internal-combustion-engine vehicle 70% of the total weight of the vehicle is iron and steel. These metals are readily recyclable. According to the United States Geological Survey recycling of automobiles is at a rate of nearly 100% each year. More than 15 million tons of steel is recycled from automobiles annually, the equivalent of approximately 12 million cars. The recycling of steel from automobiles is estimated to save the equivalent energy necessary to power 18 million homes every year.

On the other hand, for an electric vehicle only about 10% of the total weight comes from iron and steel, with almost half being due to the battery, the electric motor, and various electrical parts. Separating and sorting the materials in these components is tricky and is further complicated because of the flammable nature of current lithium-ion batteries. In Chapter 7 I described the main processes—pyrometallurgy and hydrometallurgy—that may be used to extract the metallic components in the battery. A little later in this chapter we will see that lithium is rarely one of those metals. Rather it is more usually discarded in the process of obtaining the more valuable cobalt, nickel, and manganese.

In addition to the successes seen with the recycling of iron and steel there are other metals that also enjoy high recycling rates.

Of all the metallic elements the most widely recycled is lead. In fact, lead is so efficiently recycled in the United States that over 60% of domestic lead consumption—almost all for lead-acid batteries—is satisfied by recycled material. Consequently, we are not as reliant on newly mined sources of lead as we are with some metals, for instance lithium, cobalt, and nickel the most common metals in a lithium-ion battery, as we can rely on an abundant supply of the recycled material. In 2021, data from the United States Geological Survey shows that 300,000 tons of lead were extracted, mainly from mines in Missouri and as a byproduct from two zinc mines in Alaska and two silver mines in Idaho. For the same year, more than three times as much, 990,000 tons, came from recycling. Most of this lead was from lead-acid batteries and over 90% of it will find its way back into new batteries. A cycle that can be repeated over and over again.

But although "we tend to think of recycling as an unalloyed good thing", that is not always the case especially "not the way it is done with lead in batteries."[21] Whether battery recycling is done in the United States, or Mexico, or Kenya, the toxic lead pollution can cause significant environmental and human damage. According to a joint UNICEF and Pure Earth report in many countries lead recycling is unregulated and often illegal. When the battery cases are open the acid and lead dust spill onto the ground. When smelted in open-air furnaces toxic fumes and dust enter the environment spreading contamination to surrounding neighborhoods.

In a well-regulated process, spent lead acid batteries are broken up using a shredder and the pieces fed into water-filled tanks. Because of its mass the lead sinks to the bottom of the tanks while the lighter plastics rise to the top where they can be skimmed off. The heavy metal is channeled to closed furnaces for smelting and refining before being piped into casting molds. Waste from the recycling including the acidic liquid is collected, treated, and disposed of at a designated disposal site. Everything is automated and enclosed.

In some of the un-regulated and illegal operations the batteries are drained to remove the acid then broken up using electric saws, machetes, or axes. Each of the components are separated by hand. The valuable lead is either carried or placed on open conveyor belts to take it to the furnace. Unlike the enclosed systems the furnace in some of these un-regulated operations is simply an open pot on a fire. Lead is one of the low melting temperature metals, becoming liquid at 327°C, easily within the temperature of a bonfire. Once molten the lead is poured into open cast molds where the metalworkers are exposed to the toxic fumes coming from the hot metal.

As mentioned earlier in this chapter, aluminum has an extremely large embodied energy. It is among the highest of all metals and more than ten times greater than that of lead and steel. Fortunately, aluminum enjoys a high recycling rate.

Aluminum is the most abundant metallic element comprising eight percent of the Earth's crust compared to lead which is present at only 0.0013%. So, for aluminum the reason that we recycle is not because there isn't much of it, but because of the enormous amount of energy required to separate the metal from its ore. Aluminum's main ore is the mineral bauxite, which is currently mined most extensively in Australia, China, and Guinea. Worldwide reserves of bauxite are estimated to be as much as 75 billion tons, with 24 billion tons of that total in Africa. Bauxite is converted into aluminum hydroxide by the Bayer Process invented and patented by Austrian scientist Karl Josef Bayer in 1887.[22] At the time of Bayer's innovation he was working in the Tentelev Chemical Plant in St. Petersburg, Russia. Aluminum hydroxide was an important chemical because of its use as a fixing agent —mordant—for binding dye to textile fibers.

Transforming impure bauxite-containing rock into a high-purity alumina powder is an energy intensive process. First the bauxite must be washed, crushed, and ground into a coarse powder. Then the ubiquitous silica impurities, coming from silicate rocks, are removed by dissolving the powdered bauxite in hot caustic soda (a sodium hydroxide solution) leaving behind the insoluble silica, which can be filtered off. Typical processing conditions used to dissolve bauxite are a temperature of 240°C at a pressure of about 3½ megapascals. Precipitation from the caustic soda solution produces a form of aluminum hydroxide called gibbsite. After repeatedly washing the gibbsite to remove any residual sodium it is converted into a white alumina powder by heating to a temperature of 1100°C. This process is often performed in a rotary kiln like those used for calcining cement. The average energy consumption per ton of alumina produced by the Bayer Process is around 55 gigajoules, which corresponds to around 10 tons of CO_2 equivalents per ton of alumina produced.[23]

As a point of reference, four tons of bauxite produce one ton of aluminum. More than 60,000 beverage cans can be made from this amount of metal.

Reducing the calcined alumina to metallic aluminum requires even more energy than that needed to produce alumina from bauxite. It requires lots and lots of electrical energy. In fact, it adds another 182 gigajoules to the total amount of energy required to produce 1 ton of aluminum—30 tons of CO_2 equivalents. Most of the world's aluminum is produced using an electrolytic reduction process named after its two discoverers, American Charles Martin Hall and Frenchman Paul Héroult. Independently Hall and Héroult were working on the challenging problem of how to economically produce large quantities of aluminum metal. In 1886 they both came upon the same answer, which was to become the method that is still used in industry today almost 150 years later.[24] What was key to the Hall-Héroult process is that alumina can be dissolved in molten cryolite, sodium aluminum fluoride, and then reduced to the metal by electrolytic reduction. The whole process takes place inside carbon-lined cells, or pots, at a temperature of about 1000°C. A powerful electric current passed through the bath separates the aluminum from the hot chemical solution allowing the liquid metal to be siphoned off. Each pot can produce about 230 kg of metal in a day. A single pot line may contain as many as 250 pots with a total line voltage of more than 1,000 V.

Currently, the world uses 20 million tons of aluminum each year. If all this came from freshly mined bauxite, which then had to go through both the Bayer and the Hall-Héroult processes, it would represent a staggering amount of energy. To give an idea of exactly how much energy, it would be enough electrical energy to meet the annual needs of almost 150 cities the size of Seattle, population approximately 725,000.[25] The rationale for recycling of aluminum is clear. It takes 20 times more energy to make aluminum from bauxite ore than to recycle it from scrap.

Recycling of aluminum was extensively used during the Second World War when the lightweight metal was required for the rapid expansion of the military airplane industry.[26] Between July 1940 and August 1945, the United States produced 296,000 military aircraft. At its peak American warplanes were rolling out of the manufacturing plants such as Boeing Field in Seattle at a rate of eleven every hour. More than half of these planes were made predominantly from high-strength lightweight aluminum alloys.

Although no specific numbers exist of recycling percentages during the war, "scrap drives" as they were known was a national program driven by President Roosevelt's War Production Board established in 1942. Post-war applications for aluminum, particularly in transportation, included commercial airplanes, trains, and in 1959 the iconic aluminum beverage can was introduced by the Coors Brewing Company of Golden, Colorado. An aluminum can was a new way to package both Coors's pilsner-like beer and non-alcoholic beverages, replacing the much heavier glass bottles and steel cans. It also led Coors to introduce the first recycling program for beverage cans offering consumers 1 penny for every container they returned. Between 1959 when the buy-back plan was launched and 1965, Coors collected 60 million cans. But at that time there was no process for recycling them. That situation changed a few years later when a true recycling program for the used cans was developed and they entered the production feedstock reducing the amount of new material that was needed. Data from the Aluminum Association shows that in 1970 10,000 tons of aluminum was recycled. By 1990 that number had increased by over 100 times. Since 1972 it is estimated that over 660 billion beverage cans have been recycled. Placed end-to-end they could stretch to the moon nearly 300 times. Despite the success story of aluminum recycling there is still plenty of room for improvement. In 2018 the recycling rate for aluminum cans was a little over 50%, representing 0.67 million tons.

While the recycling rates for lead and aluminum are high—over 50% globally—that is not the case for most metals. In fact, many metals that are critically important for applications in clean energy technology for instance electric vehicles, solar cells, and wind turbines are recycled at rates of less than 10% and for many including lithium the global recycling rate is less than 1%.[27]

As mentioned in Chapter 6, lithium is the essential constituent of the ubiquitous rechargeable lithium-ion battery—they are named after the metal! Lithium-ion batteries are found in every mobile device and every electric vehicle. The demand for lithium is only expected to grow because of the dramatic rise in sales of hybrid and electric vehicles. Every automobile manufacturer is now offering, or in the process of offering, vehicles powered by the movement of lithium ions. To ensure a sustainable supply of the metal for the twenty-first century it has been calculated that a 100% recycling rate with a lithium recovery of 90% would be needed. This recycling rate would be comparable to what is achieved with recycling of lead from lead-acid batteries. Unfortunately, only 3% of lithium-ion batteries are recycled and very little lithium is recovered during that process.[28] Recycling of lithium-ion batteries is far from the levels achieved with lead acid batteries. Why?

There are two main reasons lithium recycling rates are currently so low. The first is related to its availability. Although lithium occurs in the Earth's crust in a concentration of only 0.0017% it is relatively inexpensive to mine and widely distributed with Australia and Chile being the main producers. As with plastic recycling we have not yet reached the point where it is economically viable to recycle rather than continue to use primary raw materials. Most of the recycle value in a lithium-ion battery is not from lithium but from the cobalt, nickel, and manganese that are used in the cathode. When these metals are extracted, the lithium is often disposed of as a waste by product.

The second reason that lithium recycling rates are low is because both of the main recycling methods—hydrometallurgy and pyrometallurgy—have significant drawbacks. Hydrometallurgy generates high volumes of toxic and flammable process effluents that need to be treated and disposed of carefully. Pyrometallurgy or smelting is energy intensive. For 1 ton of batteries 5 gigajoules of energy are needed because of the high temperatures required. And the lithium is not even recovered. A long-term challenge to recyclers is that a process optimized for the current generation of lithium-ion batteries may not be economically viable if the battery chemistry changes significantly in the future. For instance, efforts to decrease the cobalt content could threaten the economic viability of lithium-ion battery recycling and render existing processes obsolete or ineffective.[29]

Developing efficient recycling processes and reducing our demand for raw materials are two essential steps if we are to move towards a circular economy and a more sustainable future. We can articulate our goal of sustainability using the definition of sustainable development in the United Nations Report of the World Commission on Environment and Development: Sustainable development is development that meets the needs of the present without compromising the ability of future generations to meet their own needs.[30] There is still a long way to go.

Notes

1. Geyer R, Jambeck JR, Law KL (2017) Production, use, and fate of all plastics ever made. Science Advances 3:e1700782.
2. We will use the unit "tons" to represent both the U.S. ton and the metric ton, 1 metric ton = 1.1 U.S. ton. Because of the approximation of the numbers involved in the article the 10% difference between the two weights does not impact the outcomes of the study.
3. In 1970 world production of plastics was 27 Mt. Aluminum production was 8.1 Mt and copper production was 6.1 Mt.
4. Marine Debris Program available at https://marinedebris.noaa.gov/info/patch.html.
5. Lebreton LCM, van der Zwet J, Damsteeg J-W, Slat B et al (2017) River plastic emissions to the world's oceans. Nature Communications 8:15,611. The data in this paper is presented in metric ton or tonnes. One metric ton is equal to 1.1 U.S. ton, so I have used ton to be consistent with other uses in this book.
6. Schmidt C, Krauth T, Wagner S (2017) Export of plastic debris by rivers into the sea. Environmental Science and Technology 51:12,246–12,253. A more recent model estimates that more than 1,000 rivers account for 80% of global plastic waste reaching the oceans and found small urban rivers among the most polluting, Meijer LJJ, van Emmerik T, van der Ent R, Schmidt C, Lebreton L (2021) More than 1000 rivers account for 80% of global riverine plastic emissions into the ocean. Science Advances 7:eaaz5803.
7. Woodward J, Li J, Rothwell J, Hurley R (2021) Acute riverine microplastic contamination due to avoidable releases of untreated wastewater. Nature Sustainability 4:793–802.
8. Napper IE, Davies BFR, Clifford H, Elvin S, Koldewey HJ et al. (2020) Reaching new heights in plastic pollution—preliminary findings of microplastics on Mount Everest. One Earth 3:621–630.
9. Partow H, Lacroix C, Le Floch St, Alcaro Luigi (2021) X-Press Pearly Maritime Disaster Sri Lanka. Report of the UN Environmental Advisory Mission. Available at https://postconflict.unep.ch/Sri%20Lanka/X-Press_Sri%20Lanka_UNEP_27.07.2021_s.pdf.
10. Peng Y, Wu P, Schartup AT, Zhang Y (2021) Plastic waste release caused by COVID-19 and its fate in the global ocean. Proceedings of the National Academy of Science 118:e2111530118.

11. Ashby MF (2016) Materials and Sustainable Development. Elsevier, Amsterdam, p 212.

12. The Nobel Prize in Chemistry 1963. NobelPrize.org. Nobel Prize Outreach AB 2021. Fri. 13 Aug 2021. https://www.nobelprize.org/prizes/chemistry/1963/summary/.

13. Ashby MF (2016) Materials and Sustainable Development. Elsevier, Amsterdam, pp 247, 293, 294.

14. Ministry of Environment and Food of Denmark, Environmental Protection Agency (2018) Life Cycle Assessment of Grocery Carrier Bags. Environmental Project no. 1985. Available at https://www2.mst.dk/Udgiv/publicati ons/2018/02/978-87-93614-73-4.pdf.

15. Taniguchi I, Yoshida S, Hiraga K, Miyamoto K, Kimura Y, Oda K (2019) Biodegradation of PET: current status and application aspects. ACS Catalysis 9:4089–4105.

16. Benoit N, González-Núnez R, Rodrigue D (2017) High-density polyethylene degradation followed by closed-loop recycling. Progress in Rubber, Plastics and Recycling Technology 33:17–37.

17. Bombelli P, Howe CJ, Bertocchini F (2017) Polyethylene bio-degradation by caterpillars of the wax moth Galleria mellonella. Current Biology 27:R283-R293.

18. Yoshida S, Hiraga K, Takehana T, Taniguchi I, Yamaji H et al (2016) A bacterium that degrades and assimilates poly(ethylene terephthalate. Science 351:1196–1199.

19. Celik G, Kennedy RM, Hackler RA, Ferrandon M, Tennakoon A et al (2019) Upcycling single-use polyethylene into high-quality liquid products. ACS Central Science 5:1795–1803.

20. Allison Emmerson quoted in The Guardian. https://www.theguardian.com/science/2020/apr/26/pompeii-ruins-show-that-the-romans-invented-recycling. Description of the research conducted in Pompeii is given in Chapter 4 of Emmerson ALC (2020) Life and Death in the Roman Suburb. Oxford University Press, Oxford.

21. Pearce F (2020) Getting the lead out: why battery recycling is a global health hazard. YaleEnvironment360. https://e360.yale.edu/features/getting-the-lead-out-why-battery-recycling-is-a-global-health-hazard.

22. Bayer filed British, United States, and German patents in 1887, which were issued in 1888. German Patent No. 43977 A process for the production of aluminum hydroxide was issued August 3, 1888.

23. Greenhouse gas equivalencies calculated from data provided by the United States Environmental Protection Agency available at https://www.epa.gov/ene rgy/greenhouse-gases-equivalencies-calculator-calculations-and-references.

24. Hall applied for a U.S. patent on July 9, 1886. Héroult was granted a French patent on April 23, 1886 and had applied for a U.S. patent in May. Using contemporaneous documents including two postmarked letters to his brother

George, Hall was able to prove priority in the United States. He has actually performed his first successful experiment producing aluminum from a cryolite solution on February 23, 1886.

25. Seattle City Light has about 422,810 metered customers that will use an annual 9,157,494 megawatt-hours of electricity.
26. Military aircraft production jumped from 6,000 in 1940 to 85,000 in 1943. Many of those planes were built in the Pacific Northwest where a large aluminum industry developed because of the abundant supplies of low-cost electricity.
27. Graedel TE, Allwood J, Birat J-P, Reck BK, Sibley SF et al (2011) Recycling Rates of Metals – A Status Report on the Global Metal Flows to the International Resource Panel, United Nations Environment Program. p 19.
28. Gruber PW, Medina P, Keoleian G, Kesler SE, Everson MP et al (2011) Global lithium availability. Journal of Industrial Ecology 15:760–775.
29. Mayyas A, Steward D, Mann M (2019). The case for recycling: Overview and challenges in the material supply chain for automotive Li-ion batteries. Sustainable Materials and Technologies. 19:e00087.
30. Our Common Future (1987) Report of the World Commission on Environment and Development, United Nations.

Final Thoughts

During the 100 years since the awarding of the Nobel Prize in Physics to Charles-Edouard Guillaume we have witnessed both a scientific and a technological revolution that has impacted everything from how we understand the very nature of the world around us to innovations in communication, transportation, and healthcare that touch almost every aspect of our daily lives. Over the past century developments in materials science have enabled the integrated circuit, the lithium-ion battery, the silicon solar cell, the quantum computer, the smartphone, the tools that enable manipulation of materials at the nanoscale, and among the many advances in the diagnosis of disease the now familiar rapid flow lateral test for COVID-19 that uses gold nanoparticles. In that same period, new materials have been discovered for instance graphene and high-temperature superconductors that flew in the face of conventional theories suggesting that such materials should not even be possible.

The enabling power behind these innovations has mainly come from fossil fuels: coal, oil, and natural gas. From mining of the raw materials, through purification, to eventually producing the final product enormous amounts of energy are consumed at every step of the process. In 2020 over 60% of power generation was from fossil fuels with less than 10% coming from the cleaner energy sources solar and wind. But the International Energy Agency (IEA) predicts a sea change in power generation. By 2050—midway through the current century and within a generation—the prediction is that the percentage of power coming from renewable sources could be as high as 70%.

M. G. Norton, *A Modern History of Materials*, https://doi.org/10.1007/978-3-031-23990-8

The year 2050 is particularly significant. Keeping global warming to no more than 1.5°C above pre-industrial levels, as called for in the Paris Agreement, requires emissions to be reduced by 45% by the year 2030 reaching *net zero by 2050*. Many countries around the world have committed to net-zero carbon targets, including the largest emitters, China, the United States, and India. The common goal is to cut greenhouse gas emissions that come from burning fossil fuels to as close as possible to zero. Any remaining emissions would be re-absorbed by the natural environment. A relevant question is then, "Are we on track to reach net zero targets by 2050?" The short answer is "No, we are not".

A recent report by the Intergovernmental Panel on Climate Change shows that taken together the current national climate plans for all 193 Parties to the Paris Agreement would lead to a sizable increase of almost 14% in global greenhouse gas emissions by 2030 compared to 2010 levels. If emissions are not reduced by 2030 with a continued downward trend heading toward 2050 the outcome will be the introduction of even more aggressive reduction targets to compensate for the slow start on the path towards net zero emissions. Meeting these revised targets may well come at even higher costs and present even greater technological obstacles.

The transition to a net-zero world is perhaps the greatest challenge we have ever faced. Reaching net zero will require transforming almost every aspect of our lives from what we wear, to what we make, to what we build and the way we build it. But the key to averting the worst effects of climate change is the energy sector, which accounts for around three-quarters of today's greenhouse gas emissions. Replacing polluting coal, gas and oil-fired power with energy from renewable sources such as wind and solar would dramatically reduce carbon emissions.

Meeting the International Energy Agency prediction about the dominance of wind and solar energy by 2050 brings it own set of challenges. One of which is that we could see the demand for metals such as cobalt, copper, and nickel increase by almost seven-fold. These metals and others including aluminum, lithium, silver, and zinc are critical to building an electric economy, one that does not involve burning fossil fuels. Countries with vast deposits of these and other metals will become what *The Economist* magazine calls the "green commodity superpowers" as the world transitions from fossil fuels to cleaner sources of energy.[1] These new commodities giants will vie for power potentially creating a host of geopolitical issues as poor economies and autocracies experience an economic boom with a decline in the importance of the petrostates.

The clean technologies of solar energy, wind energy, and electric vehicles depend also on large amounts of the rare earth elements. To get to net zero Europe, for instance, will require up to 26 times the amount of rare earth metals in 2050 compared with today. While the rare earth elements are not as rare as, for example lead, tin, and tungsten, only a few countries contain deposits substantial enough to mine. A sustainable future will require that we extract them with care and have a plan for what will happen to when the devices and the products that contain them come to the end of their life.

Over the past century we have created an entire class of materials—polymers—derived from crude oil. The usefulness of these materials is largely because they are indestructible and their ability to be formed into almost any imaginable shape. Unfortunately, that once extremely useful property has led to a series of unforeseen problems that affect us and our environment. From the chemicals that are used to make them to what happens—or doesn't—at the end of their useful lifecycle, plastics are becoming an increasing problem creating a pollution epidemic. But there is no slowing down or limit in our use of these materials. Plastic production is predicted to almost quadruple by 2050 and account for up to 13% of the global carbon budget. We need to design new plastics ones that can be endlessly recycled or that break down in such a way that they don't create greater environmental problems. Joshua Howgego a writer for *New Scientist* defines four keystones necessary for sustainability[2]:

Reduce - Use less material to make things
Retain - Use products for as long as possible
Recycle - Design products with their end of life in mind so they can be refurbished or the materials within them reused when the time comes
Regenerate - Where possible, use bio-based materials that break down to nourish soils when we are done with them, rather than creating long-lived, polluting waste

Sustainability will be an important consideration for the century between 2020 to 2120. It can only be achieved by moving from our current very wasteful approach in the way we extract, make, use, and dispose of materials towards a circular economy that produces little or no waste. According to many predications we will not have to wait until 2120 to determine whether our efforts at creating a net-zero carbon economy have been successful. We'll know by 2050. At that time a brief history of materials might be sub-titled "from sustainability to survivability."

Index

Printed in the United States
by Baker & Taylor Publisher Services